BLUEGRASS GRIT

BLUEGRASS GRIT

MAJOR JOSH PITCHER

DEEDS PUBLISHING | ATHENS

Published by Deeds Publishing in Athens, GA
www.deedspublishing.com

Printed in The United States of America

Cover and interior design by Deeds Publishing

ISBN 978-1-961505-39-1

Books are available in quantity for promotional or premium use.
For information, email info@deedspublishing.com.

First Edition, 2025

10 9 8 7 6 5 4 3 2 1

CONTENTS

Dedicated to Master Sergeant Joseph Kapacziewski, whose example set the bar for all the wounded Service Members to follow in remaining on active duty. Rangers Lead the Way!

To Specialist David W. Taylor from Henderson, Kentucky, who made the ultimate sacrifice for his Country and the Bluegrass State. Fury From the Sky!

To Dr. Benjamin Potter, for being the best in his field at orthopedic surgery.

To my Occupational and Physical Therapists, Anne Marie, Harvey, Bo, and Kelly, who were instrumental in returning me back to the fight.

To my Prosthetists, Mike, Art, Roger, Aaron, and Justin, true masters in their craft.

And most importantly, for my wife Michelle, who endured more than anyone I know and had the patience to support me time and time again. I love you more than you will ever understand.

Self-Pity

I never saw a wild thing sorry for itself.
A small bird will drop frozen dead from a bough
without ever having felt sorry for itself.

—David Herbert Lawrence

PROLOGUE

The improvised explosive device (IED) has become synonymous with the Global War on Terror (GWOT) since combat operations began in Afghanistan shortly after the Al Qaeda attacks on September 11, 2001. While not as prevalent in Afghanistan and Iraq during the initial phases of the operation, they soon became the most likely courses of action for insurgents to utilize in attacking United States and Coalition Service Members. They can be as simple as a hand grenade under a coffee can, or as complex as multiple anti-tank mines stacked under a pressure plate rigged to destroy a vehicle. They can be vehicle borne and driven to positions with devastating effect. They can be air delivered as kamikaze drones which are becoming prevalent on the modern battlefield. And they can be packed into boats and slammed into the sides of ships, as what occurred to the USS Cole attack in Yemen. In essence, IEDs can be any shape, color, and size. They can be employed in almost any method imaginable.

Perhaps the scariest of all though are the anti-personnel IEDs. Without specialized equipment and K-9 teams, they can be near impossible to spot. They are indiscriminate, attacking both civilian and Soldier alike. While some have triggerman who can command detonate them in order to ensure their targets are hit, the vast majority encountered, to include the one that found me, are simply dug into the earth, and left there for an unsuspecting victim. During my first tour to Afghanistan in

2012, the Paratroopers in Task Force Fury encountered IEDs of all facets. While many where found, disarmed, and cleared for the resumption of operations, some weren't. Despite the fearless and vigorous efforts of the Explosive Ordinance Disposal (EOD) teams, the K-9 teams, and those trained to clear these obstacles to allow for freedom of movement, many IEDs still remained armed and ready to inflict grievous injuries or death.

On May 6, 2011, I was commissioned as a Second Lieutenant in the U.S. Army as an Infantry Officer. On April 15, 2012, while conducting counterinsurgency operations in Afghanistan, I stepped on a pressure plate activating an anti-personnel IED. I immediately lost my left foot and much of my leg below my left knee. I had only been in the Army for less than a year.

I never imagined in a hundred years that I would chronicle my tale of a career Soldier in the United States Army. This story, set during the final surge of the Global War on Terror details how one young officer, just as ordinary as every other butter bar, had his trajectory drastically altered on a gorgeous Sunday just months after commissioning.

That catalyst, an IED in Kandahar Province, Afghanistan gave me three options. I could die then and there, which I didn't by the mercy of God. I could medically retire at the age of 23 and live off of disability or start a new occupation, which I never fathomed. Or I could disregard the fact I was now a one-legged infantryman in a two-leggers game and make the most of it, the choice I went with.

Notwithstanding my own predicament, I had to account for another dynamic. I was engaged to my now wife, Michelle, who was also starting her career. We had been together for four years and she knew absolutely nothing about the military lifestyle. Michelle now had to decide whether this was still for her. She could either walk away, unable to cope with my decision to stay in and deploy again. Or she could remain by my side, knowing that she could get another phone call in the middle of the night that her significant other was badly hurt or wounded.

Just as I decided to stick with my career, she chose to stick with me and endure it all. She bore it with a stiff upper lip, never once trying to convince me that there was honor in medically retiring.

Two questions have been inquired of me for the last twelve years. The most asked, "How did you lose your leg?" is perhaps the most straight-forward. This generally annoys me, as I am now forced to relive the worst day of my life to this curious seeker of knowledge. To civilian adults, I typically respond with "an IED in Afghanistan," and that's the end of it.

To children, I answer to their astonishment, "a bomb." They don't understand IEDs, and I don't bother explaining it to them. Their igno-rance is their blessing. It's part of my reason for fighting. To allow Amer-ican civilians to live in peace and ignorance. To other Service members though, they get a range of answers. "I loved a woman that wasn't clean," was for the enlisted infantryman. "A shark attack while surfing off of Malibu," for the other branches. And for officers who don't outrank me, they get the sarcastic "Take a wild guess." Especially for those who have never seen combat. Because those that have would never ask.

The second question I get is the most complex. "How do you contin-ue serving despite that?" By "that" I assume the missing leg. It is complex because the "how" is a result of hundreds of factors. I am not an ortho-pedic surgeon. Ask them how an amputated tibia and fibula on a human being can still be of use, despite no foot attached.

I am not an occupational therapist. Ask them how a below knee left leg amputation can be overcome in allowing the patient to complete their activities of daily living to include dressing oneself and bathing.

I am not a physical therapist. Ask them how a one-legged individ-ual can run a marathon, squat 350 pounds, or ruck for hours with fifty pounds on their back.

Finally, I am not a psychologist. Ask them how the human mind can overcome the limitations of the body and push someone to perform at unsurpassed levels never reached before.

I think you get where I am going with this. It has never been as simple as "I just do," although I wish it were that easy. Hundreds of people throughout my career have been instrumental in the infinite process of keeping me in the service. To say it is because of my resiliency is selfish. A resilient person can overcome obstacles thrown at them. But some problems require outside support. Pride will hinder even the smartest and strongest of human beings.

I could not have accomplished so much without the safety net of my team of teams who cared about me, worried about me, and dropped everything to ensure I was their priority. To these people, too numerous to list, I dedicate this work to you.

But if I can be selfish just this one time when answering your question. "How do you continue serving despite that?" My simple answer is this. "Because I choose to." I choose to because if not me, then who? I guess at the end of the day, I'd rather have to live with my wounds, both physical and the invisible mental, so someone else, who I have never met, will never have to.

Part One of this chronicle was written from the perspective of a young U.S. Army Infantry Officer at the tactical level. The thoughts and attitudes expressed in this work reflect what my eyes witnessed and what my ears heard.

National Strategies, Theater Campaign Plans, and Task Force Operations, while having a direct impact on my life, will seldom, if ever, be expressed due to my place at the lowest echelon for an officer. My experiences are the focus of this chronicle and I hope that by sharing this will assist the next generation of leaders serving their organizations, whether civilian or military alike, in any capacity.

CHAPTER 1
MAKING AN INFANTRY OFFICER
MAY TO SEPTEMBER 2011
FORT BENNING, GEORGIA

My career began on Friday, May 6, 2011 at the Douglas Whitlock Building and Auditorium at Eastern Kentucky University (EKU). The culmination of four years of undergraduate education at the "Ivy League of Appalachia," I was now eligible to be pinned with the gold bars of a Second Lieutenant (2LT) in the United States Army. The EKU Army Reserve Officer Training Corps (ROTC) Battalion had a large commissioning class this year. Along with myself, two other newly minted 2LTs, Isaac Lyons and James T (JT) Hess chose the pathway of an Infantry Officer, and with it, an all-expenses paid trip to Fort Benning[1], Georgia for the Infantry Basic Officer Leaders Course and Ranger School.

Four years had passed to the day since I graduated from John Hardin High School in Elizabethtown, Kentucky in 2007. Most of my cadre of likeminded friends chose career paths in the military too. Marty Miller followed his father's footsteps to become an enlisted Cavalry Scout

---

1. Note: Names of Fort Benning, Fort Bragg, and other installations have since changed. Until the official date of the change is reached in this chronicle, the original name will be utilized.

(19D) in the 82nd Airborne Division. Our residential genius, Drake Sullivan, was accepted into the United States Military Academy at West Point. Scott Stafford, Adam Ward, and James Masero enrolled into the University of Louisville Army ROTC Program. Although I was the only member of our clique who chose to attend EKU, we all remained close throughout our college days and enjoying the best days of our lives, not even realizing how good we had it.

My family had a long history of military service, all of it in the Army. My father, Randy Pitcher was a career Soldier who first enlisted as a Combat Medic (68W) and later commissioned as a Military Police Officer after attending Officer Candidate School. My mother, Vicki, who also served as a medic, met and married him in Germany where I was born in 1989. My father served in the storied 160th Special Operations Aviation Regiment (SOAR) and had seen combat in Desert Storm.

Randy's brother, Jeff, was an enlisted Indirect Fire Infantryman or "mortar man" (11C) who retired in 2005 and participated in the Invasion of Iraq in 2003. My own brother, Justin, was currently serving as an Infantryman (11B) in the 2nd Infantry Division at Fort Lewis, Washington. Needless to say, it was in my blood to serve my country. At no point did I ever consider any other occupation.

From 2007 until 2011, I was a Cadet in the Colonel's Battalion where I learned the ropes of Army life. Cadet life came naturally for me as I was in great shape, grew up in the woods, and was accustomed to the protocols of the Army community through a childhood living on military installations throughout the United States. The Cadets lived on the 10th and 11th floor of Todd Hall and it would be an understatement to say we were a tight knit group.

Apart from physical training at 0600 hours three times a week and some outdoor training every Thursday, the life as a Cadet did not diminish my ability to enjoy the freedoms of being a college student. I joined the Lambda Chi Alpha fraternity during my first semester and made some outstanding friends that helped me break out of my awkward high school shell. Spring breaks at Panama City, South Padre Island, and San Diego were executed where I continually tested the limits of my soul and liver with friends. My biggest companions at EKU were Nick Cory, Josh Cole, John Hoekzema, Josh Bailey, Anthony Moore, and Crosby Kennedy. Shooting guns, drinking Natural Light, and floating down Madison County's Silver Creek on inner tubes on hot summer days with a handle of Evan Williams could best characterize our free lifestyles.

To be fair though, despite having a very high-grade point average, I wasn't exactly on a righteous path as I partied hard during the day and enjoyed the companionship of beautiful students in the evenings during my freshman year. I had more than my fair share of run ins with rival frat boys or the law and just wanted to continue living my life like a free bird before the constraints of life as a military officer took hold. Then, in September of 2008, my life changed in the most extraordinary fashion when I did something I swore I would not do. I fell in love.

On the 9th of September 2008, at roughly the start of my sophomore year, two fellow Cadets, Isaac and Andrew Chinn attended a mixer hosted by my fraternity for the Deta Zeta Sorority. Over the summer, my girlfriend and I went our separate ways. I was now ready to play the field again. At a house on Robin Way, about one mile from campus, about fifty students were burning the place down in a party that perfectly started the year. That's when it happened.

I looked over from the bar towards the flip cup table. A tall, slender, and gorgeous brunette was keeping up with the boys in a heated match where Andrew was competing on an opposite team. After the match, Andrew walked over to the bar to grab a couple of red solo cups filled Bud Lights. Then I locked eyes with her for a second. She had this wicked smile about her and as soon as we stared into each other, she blushed and turned away. I had this one chance to take her home before some other lucky man made the move.

I walked over and immediately introduced myself and we quickly struck up a conversation. Michelle Lynn Smith was her name, and she too was in her second year at EKU, studying social work. Andrew was walking over with the beers when he saw us talking. "Damnit!" he exclaimed and walked away in defeat.

For what seemed like hours that night, we talked about our hobbies, interests, and about our lives. I was smitten by her instantly. Later that night, we did go home together, but the angel in her kept me from pursuing any other motives that were unchaste that evening. Within two days, we were dating. That was the last time I have ever been single.

For three years, the two of us were inseparable. We practically lived in each other's dorms until we both started renting with friends off campus. We went on spring breaks together and I was attending church with her parents, Vicki and Clay, with whom I came to love as my own. After a year dating, I knew she was the one. Something in my soul just told me, "She will love you unconditionally and will be unconditionally loyal," the perfect formula for a military spouse.

I proposed to her on top of the cliffs overlooking the Bluegrass known as the Pinnacles in Berea in the fall of 2009. Her engagement ring was my first major purchase, but the investment has paid itself back infinitely.

I was on top of the world in the spring of 2011. I was engaged to the love of my life. I was graduating with a B.A. in History with zero debt. And I was following my family's footsteps by joining their ranks through service towards the greatest nation on Earth. When my parents and fiancée pinned my 2LT boards on my shoulders that sunny afternoon, it was my biggest accomplishment to date and the look of how proud my parents were of me has stayed with me to this date.

But my celebrations were short lived, however, as I reported to Bravo Company Class 7-11 IBOLC (Infantry Basic Officers Leadership Course – for our Vietnam era readers, this was called IOBC – Infantry Officers Basic Course in your day) the following week. I had exactly four days to pack, move, and settle in. Not exactly enough time to go on a vacation or enjoy some well-deserved fun.

JT and I were in the same class while Isaac would attend training later that summer in a different class. Saying goodbye to Michelle who had another year until graduation, I set off on I-75 south for Dixie. Despite four years of preparation, I had absolutely no idea what I was in for.

Fort Benning, Georgia straddles the Georgia-Alabama border in the vicinity of Columbus. Home to the U.S. Army Maneuver Center of Excellence, IBOLC, infantry basic training, and the United States Army National Guard's Warrior Training Center, the installation is teeming with new recruits and newly minted officers. In addition to these programs, Ranger School and Airborne School call Fort Benning home. The vaunted 75th Ranger Regiment bases its 3rd Battalion and Headquarters here as well.

A year ago, I had attended the Basic Airborne Course at Fort Benning after my ROTC culminating exercise called Leadership, Development, and Assessment Course (LDAC) at Fort Lewis. Apart from the training, which taught me how to jump out of a perfectly good airplane and land, I remember the experience being incredibly hot and humid.

Taking the advice of some previous EKU graduates who had completed IBOLC, I decided to live off post in a small apartment community immediately adjacent to the Sand Hill gate. Independence Place, known affectionately by many 2LTs as Forward Operating Base (FOB) "FOB IP" provided furnished apartments where you could either live single, with a buddy, or three others. To a new officer with no furniture and new to the area, it was a great way to meet new people and avoid spending money on furniture.

I began in processing into IBOLC on the first duty day after arriving in Columbus, GA. The class officially started the following Monday, so the majority of my first week consisted mainly of attending an 0900 hours formation on Taylor Field and providing a document when requested. The majority of the in processing would begin the subsequent week when all my medical files were reviewed. I conducted numerous medical screenings and conducted various administrative actions to initiate my pay and benefits.

IBOLC Class 7-11 consisted of Bravo Company which was divided into four platoons of around forty students per platoon. Almost all the students originated from OCS or ROTC as West Point had not completed graduation this early in the year. I was assigned to Third Platoon or "Third Herd" which was subsequently broken down into four squads by alphabetical order. JT was in Fourth Platoon, so I was on my own from here on out.

The first questions going around consisted of where the members of the platoon where living or where they hailed from. A good quarter of the group also lived at FOB IP and we quickly struck up conversations, regardless of what squad we were in. I immediately hit it off with a prior service 2LT from Minnesota named Adam Lawson. A veteran of the Invasion of Iraq and the 3rd Brigade Combat Team of the 101st Airborne Division "Rakkasans," Adam was also engaged to his fiancée Lauren who remained back home. Others who quickly enlarged our group were 2LT Hank Gray, 2LT Charlie Fulton, 2LT Ean Pokryfky, 2LT Nick Rich, 2LT Connor Rouse, and 2LT Dustin Lawrence.

IBOLC in 2011 was 16 weeks in duration and consisted of a solid 50-50 mix of training in the woods and classroom instruction. One week was spent in the summer heat either conducting marksmanship ranges, tactical patrol lanes, maneuver live fire training, or land navigation and forced marches. The subsequent week was spent in the air-conditioned classrooms adjacent to Taylor Field where we learned how to develop operations orders, principles of offensive and defensive operations, and honing our briefing skills. The week on and week off construct kept many students from burning out and gave us an adequate amount of time to finish assignments and refit our equipment after a week in the Georgia wilderness.

Within a couple of weeks of the course, we met our senior instructor who had recently graduated from the U.S. Army Pathfinder Course. CPT Michael Linton introduced himself to Third Herd one Monday morning at 0600 hours where he subsequently ran the entire platoon two miles from Taylor Field to Godman Army Airfield. A famous incline, called Cardiac Hill by all students, led from the main base cantonment area to the airfield and took the breath out of even the best runners.

Once all members of the platoon had reached the bottom, we knew we had to turn around and make our way back up. But CPT Linton had other plans. "Buddy up," he ordered. I found a partner in 2LT Kevin Nichols, a phenomenal athlete with a deadpan sense of humor. "One of you is wounded at all times and cannot walk," CPT Linton proclaimed. "Return to Taylor Field and don't cheat," he finished before taking off.

A chorus of profanity under the breaths of forty officers followed along with quick changes in partners for those who chose poorly. Nichols weighed at least 15 pounds more than me, but his endurance would help greatly in getting us back to the finish line. For over a half hour, we took turns fireman's carrying each other. Although it was still early in the morning, the temperature was already over 90 degrees.

The first obstacle to us was Cardiac Hill which challenged our calves and hip flexors to every degree. We switched out every hundred meters to conserve our strength, but at around the halfway point, it no longer mattered.

Two couples were ahead of us while over 15 other pairs lagged behind what looked like a macabre trail of officers questioning their life's choices. Thinking my legs would give out from under me, 2LT Nichols and I arrived at Taylor Field as the third pair. We waited for the remainder of the platoon to finish, the designated time for PT having been completed well over a half hour ago. Now our shower and breakfast time was being cut into.

This was our introduction to CPT Linton, who became our bane during the first half of the course. But his lesson was simple. Be in the best shape ever for the toughest day of ground combat. Our bodies, in the prime of our youth, still needed much more work on strength, conditioning, and endurance in order to rise to the occasion when needed. Strong legs could make the difference in any situation, As an officer expected to one day lead a platoon of infantryman, strong legs were necessary. Though I was a stud at the age of 22 and was stronger and faster than most, I still needed work.

Ranger School was to be the litmus test for most of the class, and CPT Linton set about getting us ready for it.

The highlight of IBOLC was the weekend. Weekends and four-day pass's during the Memorial Day and Fourth of July weekends were what we lived for. Every Friday afternoon after final formation, the gang met at the pool with a bottle of Jamison and case of beer. For hours we marinated in the pool until rallying behind a designated driver for dinner at the Vallarta Mexican Restaurant outside Victory Gate. After consuming my weight in tacos and Modelo, our group took a trip to Broadway Street where the entertainment of the venues The Cannon and Scruffy Murphy's kept us going until early the next day.

Adam, Hank, and I called our group the Friday Night Crew or FNC and these nights were some of the finest times of our lives. Avoiding the Fort Benning Courtesy Patrol, picking fights with Ranger "Bat" gents who would surely have kicked our asses, or making complete fools of ourselves by the copious amount of alcohol consumed, we always made it back to FOB IP or our respective accommodations through the intense planning for who the DD was during the week's training. A DUI was a surefire way to be removed from the course and a fast way to end your career before it even got off the ground. Luckily for our platoon, we only had one DUI during the entire 16 weeks, not a member of FNC but a loss nonetheless to our team.

As fate would have it, I had a target on my back from the good CPT Linton. From extra PT to causing the platoon to low crawl for a hundred meters through fields riddled with fire ants, I couldn't do or say anything without incurring CPT Linton's wrath and initiating collective punishment for the whole team. Then, in the middle of the course, I was selected to become the Bravo Company Student First Sergeant (1SG). In addition to my role in Third Herd, I now had to lead the entire company in formations, administrative actions, and delegating platoons for duties such as classroom cleanup or serving chow in the field.

To make matters even more interesting, a draconian Scottish Sergeant Major (SGM) on loan from the British Army was the Senior Enlisted Advisor for the company and we absolutely were petrified of getting on his bad side, which coincidently I was always on.

For over a month, I endured getting blasted by SGM Andy Lambert in front of the whole company for my perceived failures to not knowing why somebody was late, missing, or how somebody had not secured their trash. I do not recall the last guy having it this bad. Hell, due to an administrative error on the Army's end, I wasn't even paid for my first three months and was beginning to struggle financially. However, after the lickings were delivered, my platoon was there, to welcome me back with open arms and remind me that it would soon be over and another unlucky 2LT would take over for me.

By the Fourth of July weekend, three miracles occurred, I was finally out as the class 1SG, CPT Linton was transferred to a new assignment and replaced by CPT Kyle Tarvin, and my pay had finally hit my bank account. Two months back pay for a new 2LT doesn't seem like much now, but for a 22-year-old straight out of college, it was like hitting a small lottery.

Unfortunately, Uncle Sam took his due in the form of taxes so it was significantly less than expected. For the longest time, Adam had been loaning me petty cash to get by, which I promptly repaid in full. I hate depending on others for anything and the humility I had to experience by accepting cash to pay rent was a hit on my ego. As I would learn a year later, sometimes you absolutely need to accept assistance in order to get back on top.

Halfway through with IBOLC, I learned how to qualify on the M4A1 Carbine, honed my land navigation skills, qualified in Level One Combatives, and was a certified Combat Life Saver after a weeklong first aid course. I had executed live fire training as a buddy team, a fire team, and as a squad and had gotten the knack of patrolling as a member of a platoon.

Most importantly as an officer, I had planned and briefed incredibly detailed operations orders from dismounted platoon attacks to mechanized operations in an urban environment. Every morning was an intense PT regimen ranging from running or rucking for miles or a CrossFit workout involving body weight exercises. I was not only getting into the best shape of my life, but was gaining the confidence and competence necessary to succeed as a Platoon Leader, the destiny of most IBOLC graduates.

During this time, I missed Michelle immensely. I guess I knew I truly loved her when, after a brief separation that summer, I called her just to hear her voice. It was as if nothing had happened at all while she described finding and naming a baby rabbit she had found on her parents' property. While some may scoff at this as childish, that was one of the quirks about Michelle that has always drawn me to her. A lover of nature and anything furry, she is one of the kindest and gentlest souls one could meet, and I did not know yet just how fiercely loyal she could be. Realizing that I wanted to press on with the engagement, we immediately reconciled.

In August, my high school friend Drake Sullivan joined me at FOB IP, having commissioned as an Infantry 2LT from West Point. Isaac Lyons, one of my roommates from EKU also began his training with the next class. Unfortunately, our time for reunions and catching up on beers was ending. I was wrapping up my final weeks at IBOLC and preparing to begin my final training prior to conducting a Permanent Change of Station (PCS) to Fort Bragg.

All my class had remaining was a final operations order exam, a twelve-mile road march, and a platoon live fire exercise. Graduation on the 9th of September was on everyone's minds as it meant we could officially don the blue cord of the Infantry on our Army Service Uniform (ASU).

Immediately after graduation, I was enrolling in Ranger School which would begin only three days after completing IBOLC.

Ranger School, one of the toughest and finest courses in small unit tactics that the U.S. Army has ever devised was to be my fate in September. The culmination point for infantry officer training, IBOLC's curriculum really set the officers up for success in understanding and grasping the concepts of fighting as a squad and platoon that made up the bulk of Ranger School. Consisting of 61 total days of training, the first 10 days of the course weeded out the unprepared in physical fitness exercises, land navigation, obstacle courses, forced marches, and water survival. The rest of the course was divided into three phases, Benning Phase, Mountain Phase, and Swamp Phase. Each phase consisted of small unit tactics where students are graded on their competence, leadership abilities, and mental toughness under pressure.

To move on to the next phase of the course, a student must pass his patrol.[2] As students' progress to each subsequent phase, the patrols get longer and the missions more difficult. Students who failed to achieve a "GO" would be recycled from the phase into a next class and for some underperformers or those needing more training, a day one restart could be recommended in lieu of being dropped from the course. The 61-day course could now become 120 days or more if a student kept being recycled.

The last month of IBOLC consisted of diagnostic and assessment physical fitness tests and physicals for Ranger School. The cadre did everything in their power to prepare us but when it came down to it, only those who really poured their heart and souls in their preparation and study would ever be ready for Ranger School. The packing list alone for the course was vast, over ten pages of fine print of materials, equipment, and clothing needed just to enter the course.

Back in the day at Fort Benning, you had two options, Commando's

---

2. Note: during this era, only men could attend Ranger School.

Supply or Ranger Joes. I went with Commando's due to a referral from a friend who informed me that they already had the packing list for the course ready for pickup. All you had to do was inventory the contents from a giant bag, check off a copy of your packing list, and pay for the items. It couldn't have been any simpler and the packing list layout we were told was one of the first tests of earning the Ranger Tab on the very first day of the course.

During this time, my brother deployed with his unit to the same province in Afghanistan where my future unit would operate. An Automatic Rifleman in the 5th Stryker Battalion of the 20th Infantry Regiment, 2nd Infantry Regiment, he would have our families first baptism of fire in nearly a decade. From what I would later learn, his outfit endured a lot of combat from the Taliban in their Area of Operations (AO) and incurred their fair share of casualties from IED's and small arms fire. I was proud of him and used his tour of duty as fuel to motivate me to get through this next challenge.

One final surprise was in store for me during my final week in IBOLC. For reasons unknown to me, CPT Tarvin ranked me number one out of the forty other officers in Third Herd. That meant I now had to represent the platoon in a board where I competed against the other three top officers for the Class 7-11 Distinguished Honor Graduate. I was somewhat taken aback by the honor as I could have named dozens of officers that I felt merited the title of Platoon Honor Graduate. Nonetheless, CPT Tarvin's decision was made on a holistic level incorporating all aspects of my performance, academics, and peers throughout the course.

Two days before graduation, I stood before a panel of officers who were the Class 7-11 tactical trainers; the Company Commander, CPT Baldwin; and SGM Lambert. For an hour I was interrogated on my knowledge about what I learned in the course. "What are the principles of patrolling?" "What does TTLODAC stand for?" "What is your favorite book and why?" were examples of some of the questions asked.

Then it happened. I was asked, "Why do you think you are the top student in Class 7-11?" My answer was straightforward and bold. "Well Sir, I am the top student because I had the best instructor, it's as easy as that." The Company Commander bellowed, "So you are saying CPT Tarvin is better than the other instructors on this panel?" "Absolutely, Sir," I replied with confidence. "Get out of here!" I was ordered to leave the office and wait outside with the four others.

I didn't have to wait long as I had been the third candidate to board out of four. After another hour, we were all called back inside and lined up abreast to the panel to await their verdict on the winner. "2LT Pitcher congratulations, you have been selected as the Distinguished Honor Graduate for IBOLC Class 7-11," the commander informed me.

I shook hands with the board panels and the other candidates and made my way outside to Taylor Field where the gang was busy cleaning weapons for the sixth time prior to turn in. "How did it go?" Adam asked. "Did you win?" said Hank. "Yes, I won," I said, downplaying the achievement. To a hail of congratulations, my half-cleaned weapon was passed back to me along with my weapons cleaning kit which had been on loan during the board. Never again would cleaning my weapon be so easy.

Graduation came and went on September 9, 2011. I had been in the Army now for four months and had already accomplished much. The real challenge though, began in 72 hours. That weekend, Adam and I spent most of our time checking over our packing lists and getting our hair cut for zero day that following Monday. A rite of passage for all students, shaving your head for Ranger School had healthful and pragmatic benefits as we would have no time to comb our hair or shower but maybe once a week. Since Adam was already a decade older than me, his hair was already thinner and thus, it didn't matter much to him anyway.

Finally, on a beautiful early fall day, we locked up our rooms at FOB IP and dragged our duffel bags and ruck sacks to the parking lot. Adam pulled up in his small pickup and we filled the bed with about five oversized bags bursting with the mandated packing list. As we drove to Camp Rogers on the outskirts of Fort Benning, a sense of excitement and wonderment overcame me. This was it. The final test prior to donning the Ranger Tab, taking a platoon, and wearing the coveted ranger panties that were all the rage in Columbus, Georgia during the summer.

My wonder quickly dissipated though as I entered a land of barbed wire and black and gold structures. "Not for the weak or faint hearted." a sign near the entrance proclaimed. To many of us, it must have read more like "Abandon all hope, Yee who enter." "Well, here goes nothing," I must have told Adam. "How bad could it really be?"

Author and father, SFC Randall Pitcher prior to deployment for
Operation Desert Shield/ Storm, (1990)

LEFT: Cadet Josh Pitcher with his Fiancé Michelle Lynn Smith just one month prior to Commissioning from Eastern Kentucky University (MAR 2011). **RIGHT:** Josh Pitcher in 2010 in his lifelong passion of hiking. Pinnacles, Berea, KY (SEP 2010)

2LT Pitcher with his IBOLC friends during a break from training (JUL 2011)

CHAPTER 2
CAMP ROGERS
SEPTEMBER 2011
FORT BENNING, GA

Day Zero is the only easy day of Ranger School, apart from graduation. Consisting of in-processing and equipment layouts, there is no yelling or physical fitness punishments commonly known as "getting smoked." 344 students made up our Ranger Class. On a large open area where hundreds of students were huddled with their baggage, NCOs wearing army combat uniform (ACU) trousers and black t shirts with a large black and gold Ranger Tab on the front calmly began the process of separating the mass into three companies. I was placed into Bravo Company and instructed to move with my bags with haste to a large gravel pit under a pavilion about 100 meters away.

Although I was not yet being tested, I was still stressed with anticipation about the unknowns that lay ahead. As I moved as quickly as possible with my ruck on my back and carrying a duffle bag weighing close to 100 pounds in each hand, fears of being singled for any trivial infraction kept my eyes down and legs moving. My anxiety quickly dissipated though as I arrived at the awning and heard students joking and talking.

A familiar voice called from the gaggle, "Pitcher!" It was JT Hess! A familiar face in my company. Adam Lawson was there too, talking to another IBOLC graduate. Dropping my bags next to my feet and wiping a drip of sweat from my brow, I quickly joined the same conversations everyone was having. "I hope we are in the same platoon." "Do you think they'll began smoking us today to wear us down before the PT test?" "Where is your unit?"

Most of the company came from all facets of the Army and was comprised of a healthy mix of enlisted and officers of all ranks, MOS, and units. IBOLC, the 75th Ranger Regiment, and Soldiers from the 101st and 82nd Airborne provided most of the numbers for this class, I quickly realized. It would be smart to make friends with everyone and not alienate a soul. Piss off one "bat boy" and you'll feel the wrath of his teammates, I immediately contemplated.

One of the black t shirt NCOs walked up to the pit. Ranger Instructors (RI) are the cadre who lead training for all Ranger School students. Everything we did was under their eyes and supervision. In short, for the next 61 days, they were our lord and masters and determined our fates. Approaching the gaggle, the RI did not look pleased. In a southern drawl he barked, "Aren't we in the Army? Get into a formation now and be prepared to move into your designated platoons." A hundred Soldiers quickly grabbed their bags, and a rectangular mass began to take shape. After three minutes, we stood in silence. "Close enough," the RI muttered. I heard another student say, "Tomorrow will be different, we'll have two minutes and when we fail, we'll pay for it." I later learned he was a day one recycle and was starting over completely. He knew what to expect.

Names were called off and roster numbers disseminated. Three small rectangles began to take shape. I was placed in Third Platoon, Third Squad and given the number 257. For the next two months, I was no longer 2LT Josh Pitcher. My uniform had an American flag on the right shoulder, U.S. Army tape, and my name tape. No unit insignia, no badges, no rank. I was now 257 to anyone outside of my squad.

Comprised of twelve students, the group had three Rangers from the 75th Ranger Regiment, two IBOLC students, two NCOs from the 82nd Airborne Division, a foreign student from Lebanon, and four junior enlisted infantrymen from the active component. By a miracle, the only IBOLC graduate was none other than JT Hess! At least we knew each other. Adam was in Second Squad and looked as calm and collected as ever. Two friends in the same platoon. What luck!

The next task was the packing list layout, which was our final action that day. Since this was Day Zero, Ranger School didn't officially begin until tomorrow after the first test of the Ranger Assessment Phase commonly known as RAP Week. Bravo Company's platoons each occupied their own pavilion and gravel pit. By platoon, almost fifty shaven students were ordered to spread out equally in the pit in order to maximize the space required for the RIs to inventory every article on the packing list. A small team of RIs stood in the front of the group. Their leader, a SFC with the same Georgian drawl barked out a set of commands. "Dump out everything and spread it out. Every item is to be taken out of a bag or its container for inspection. You have five minutes." Eyes fixed on the RI, we waited for the command. "Begin!"

Students emptied the contents of their duffle bags and rucks into a giant pile in front of them. A cascade of noise occupied the pavilion as students removed items of clothing from carefully sealed bags. Organization evaporated in the blink of an eye as socks, underwear, shaving articles, and weapons cleaning kits were dumped out onto the dusty gravel for inspection. Some poor students, who had gone through the trouble of vacuum sealing their clothes to save space, struggled to unseal every article of clothing.

"Time!" the RI bellowed. "Sit down if you're done." Out of fifty students, only half had completed the task. "If you can't move with a purpose here then you won't last long," an RI informed us. "But since you need a little motivation, let us help you. Front leaning rest position, move."

So, it came to pass now. The rumor of no physical punishment prior to the PT test was but mere gossip. "In cadence, exercise!" In unison, the platoon lowered their bodies parallel to the ground. Each one of us yelled "one…two…three…one" as we performed the performance measures to the pushup at the RI's slow command. After the tenth repetition. We held our place. "Recover!" was given. The entire platoon rose together to the position of attention and belted out "Ranger!" While ten pushups aren't anything to a fresh infantryman, this was just a harbinger of what lay ahead of us. That wasn't a smoking. That was but a taste. A warning.

"Two minutes," the senior RI commanded. As the cacophony of movements from Soldiers struggling to finish the layout commenced again, I looked around at the madness. A vast landscape of clothes, equipment, and sundry items filled the gravel pit. Individual piles no longer existed but blended into each other's packing list layout like a congealed mass. It was controlled chaos by the RIs. "Time! Sit down if complete." By a miracle, everyone either sat or squatted over their contents. "Good then," the RI reflected. "Now let us begin."

A Sergeant First Class (SFC) RI read off the packing list item by item. When he called out the item, every student had to hold it up for another RI to acknowledge. RIs moved like sharks throughout the mass, stepping on equipment and clothes. If a student was missing an item, his roster number was taken up after a brief chastising. For over an hour, we executed this evolution until the entirety of the inventory was satisfied.

We then were given five minutes to pack up, which, just like the layout dump, many of us failed. The result was more pushups and reminders about how pathetic we were as a platoon. My name was not called though, the Commando's packing list purchase had paid off. Carrying all of my gear back to my locker in the barrack, I tried to quickly rack out. We had only six hours till wakeup and that would be the most sleep we were given for a long time.

I rose from the bottom of a double bunk bed at 0300 hours. I slept in my PT uniform that night and had shaved that evening to save time.

This was now day one of Ranger School. At approximately 0500, the PT test would begin as the initiation of our class into RAP week. Consisting of two minutes of pushups, two minutes of sit-ups, a five-mile run, and seven dead hang chin ups, the Ranger Physical Fitness Test (RPFT) tested our baseline fitness prior to any student entering the course. Traditionally, the RPFT sends more students home for failure than any other event besides land navigation during RAP Week.

Breakfast was not served this morning. Business first. The company was assembled at 0330 and readied to move to the knife fighting pit by the Malvesti Obstacle Course immediately outside the gates of Camp Rogers. Over 340 students marched in their platoons to the pit in silence. There were no cadences or noise. We didn't need to sing to be motivated. Either you wanted the tab, or you didn't.

By 0400 hours, the entire class was assembled around a large contingent of RIs. An RI read off the task, conditions, and standard for each event of the RPFT and a demonstration was given for the stationary events. Then, we were ordered to form ten lines towards the front of the pit. I found myself at number four in a line. I wouldn't be first, and I sure as hell wouldn't be last.

The entire class was now intermixed throughout the ten lines of students. Company and platoon integrity vanished, and the bald masses of students now waited for their turn. "First Rangers come forward," the NCOIC (Non-Commissioned Officer in Charge) of the team of RIs commanded. "The rest of you, about face." Over 300 students turned around with their backs facing the knife fighting pit where the pushup and sit-up event of the RPFT was being conducted. This purpose to prevent the students from observing the test and trying to determine how the pushups and sit-ups were going to graded. "Get ready," the RI said, the signal for the students being graded to assume the front leaning rest. "Begin."

The only sounds heard were the RIs grading and the exertions of the students. "One minute left." Facing the Malvesti Obstacle Course opposite the pit, I could hear some RIs giving feedback to the students performing their repetitions. "Not low enough." "Parallel to the marching surface." "I'm not going to count those, Ranger." The RIs were not about to give out freebies on what would be the most important PT test for many of these Soldiers' careers.

"Thirty seconds remaining." The sounds of students struggling to eke out a correct repetition intensified. As the countdown from ten began, audible struggling soon gave way to despair to some. "3…2…1, halt." The students who passed were then ordered into another ten lines. The ones who didn't were isolated into their own group. Some students attempted to plea their cases with their graders, but to no avail. After two minutes of attempting to perform the passing count of 49 pushups, they were smoked. Some of the students performed at least 100 pushups by the looks on their faces. But at least 52 of them were not to the standard of the RIs.

"Next ten forward." For the next three groups, it was a repeat of the first group. I began to grow slightly nervous as there was now nothing in my way from my two minutes in the spotlight. "Next group up." I turned around and ran to the closet black t-shirt to me. He quickly told me, "Okay Ranger, 49 correct repetitions. This is nothing new."

"Roger Sergeant," I answered with as much confidence as I could muster. I dropped onto my knee's perpendicular to my grader. Eyes forward, I ignored the settings around me and focused on what task needed to be done. "Get ready!" I assumed the front leaning rest and went as stiff as a board. I had a plan. Slow repetitions with exaggerated movements showing that I was locking out my arms and lowering myself past the point where my shoulder blades and elbows aligned. I wanted my grader to know I was taking this seriously.

"Begin!" I lowered myself and paused for a microsecond. Then I lifted my body until my arms were fully extended. I kept my eyes forward and my back as straight as possible. My RI did not give me any confirmation that the repetition counted. So, I performed another pushup, with the same degree of concentration. After 10 repetitions. I heard a faint "ten." Okay. He was counting them. My confidence restored, I set to the task at hand. At around the call for "one minute" I reached my goal. I glanced at the RI and to my shock heard, "Okay Ranger, do one more for good measure." I performed one more and waited for the signal to recover. "Okay Ranger, recover and move to the lines over there. Turn around and don't talk to anyone." My first task at Ranger School was complete. Now for the sit-ups, pullups, and the five-mile run.

The remaining events passed in a blur. After 59 sit-ups, seven pullups, and a nice brisk five mile run in 32:30 minutes, I joined a mass of students waiting for further instructions. I looked around to see students still performing their pushups or sit-ups. These were the students retesting for a failed event. To what I recall, almost none of these students passed. This group was corralled away from the students still in the game.

I noticed a familiar face in the group. Adam! "Damn," I muttered under my breath. I was going to miss my friend who had kept me going during IBOLC. Out of the dejected faces in his company, he seemed to be the only one with his head held high. Bastard gave it the old college try I reckoned.

Now that the RPFT was over, RAP week was in full swing. Unfortunately, the coolness of the RIs immediately soured. For every offense, real or imagined, the students were smoked incessantly. Not making time hacks, forgetting pieces of equipment, misplacing a vow in the Ranger Creed, or simply telling a bad joke was now greeted with various exercises designed to attack joints and muscles from every angle. Between getting smoked and moving as a formation to every place, we now performed the events of RAP week.

Night into day navigation began the next day at 0400 hours. For three hours I scoured the woods east of Camp Rogers looking for a metal point. I had always enjoyed land navigation. Just you and God in the woods. There is nobody else to distract or slow you up. I have a simple strategy. When I reach where I believe my point is, I stop and sit down.

I take my cap off and just listen as if I was waiting in my tree stand for a buck to wander by. Sure, enough though, I hear the rustle of leaves and sticks breaking beneath feet.

Then, I see a red light illicitly being activated and waved around like a ship in distress. I watch patiently. "Hell yeah!" I hear as a student about 15 meters away moves at an angle from me to a red box perched on a stake. With his light on, he writes the code on the box and punches his scoresheet. As he moves away, I move in and immediately annotate my card and physically mark it as proof of my visit. Then, I disappear and move on to my next point.

I had no issues that morning finding my five points. As the beginning morning nautical twilight (BMNT) began to illuminate the woods, I moved quicker. I used trail intersections as attack points to cut down on the pace counting. When I had enough points to pass the evolution, I looked at my watch. With over an hour to spare, I traveled cross country to my last point before sprinting to the turn in station.

With thirty minutes to spare, I moved to a table in the middle of the woods where the class had been bivouacking. I turned in my scorecard and held my breath as an RI graded my sheet and ensured I had all of my required equipment.

"What unit are you going to Ranger?" the RI asked, as though he already knew I was an IBOLC grad. "The 82nd Airborne, Sergeant!" I stuttered out, already feeling the effects of the last two days. "You'll do fine there," he told me. "Go over to your platoon area and rest, Ranger."

I breathed out a sigh of relief and felt lighter than air as I dropped my ruck to the earth and relaxed. My second task for RAP week was secured. I was one step closer to making the cut and having my ticket punched for the next phase of the course at Camp Darby.

The class was getting smaller now. Less than 300 remained of the original 344 who began. And we had only been training for two days, out of 61. The next event was the Combat Water Survival Test at Victory Pond. Consisting of four events, this evolution is designed to see if you can swim and be trusted to overcome your fears of heights and the water. In essence, are you going to panic in water is what the instructors want to find out.

Students started with a short drop off the ledge into the water where they ditched their equipment. The student then had to swim 20 meters to a ladder and climb out. The uniform is your combat fatigues and boots with a rubber rifle. Next is a ladder climb up to about forty feet where students had to confidently walk across a beam measuring about one foot in width with 20 meters of distance to travel. In the center of the beam are two steps students had to cross over.

For the faint of heart or those with a fear of heights, it is indeed a daunting task. At the end of the walk, students shimmied on a rope and for about ten meters, pulled themselves to a suspended Ranger Tab and then hung by their hands. "Permission to drop!" the student pleaded. "Drop, Ranger," an observing RI commanded. Letting go of the rope, students sounded off with "Rangers Lead the Way!" and plummeted into the water below.

The final test was by no means difficult, but rather fun. Ascending on a staircase to a height that must have been at least 80 feet, students rode a zipline into the depths below. Hanging by just my hands, I looked ahead at a traffic light mounted towards the end of the zipline. Once the color changed from red to green, you were commanded to let go and fall into the water. Once complete, the water challenges of the day were over.

With three events behind me, only five more events remained. That afternoon was the 2-mile buddy run and the Malvesti Obstacle Course. While the run was simple, the Malvesti O Course was a nightmare. Consisting of obstacles like monkey bars, ladders, and nets, the course itself is nothing too strenuous. However, with the turn of a facet, RIs flood the field and the red Georgia clay turns into a quagmire of red mud up to your knees. To compound the mud and the obstacles, RIs smoke the hell out of you between obstacles while you wait your turn to execute.

By the end of the course, you're physically spent. I remember my uniform being ruined from the orange muck. Along with several other students, I promptly disposed of them as is, not even attempting to clean them.

The next day was Ranger Stakes, an exam where students proved they could perform basic Soldier tasks such as placing weapons into operation or programming radios. Then there was a 12-mile ruck march with 35 pounds. JT and I paired together for the two six-mile loops around the compound. We made good time and were never concerned about whether or not we would finish.

Finally, students were bussed to Camp Darby where they finished their RAP Week tasks with the execution of the infamous Darby Queen Obstacle Course, one of the hardest courses in the Army and over a mile long. J.T. Hess and I stood next to each other in a line, one of the first pairs to attempt the Darby Queen. As with everything Hess had attempted at EKU, he naturally crushed the different obstacles. Having him as a partner made it not only fun but competitive. In the end, we made it through all obstacles on our first try.

In the end, the class had shrunk significantly over the course of a week. Between failing events, injuries, or quitting, we were hemorrhaging numbers. The attrition rate was nothing new though. It was just another day in Ranger School for the RIs.

The platoons were reformed at the end of the week to ensure equal displacement throughout the squads. Hess was still with me in the squad. With RAP Week behind us, the students had passed their job interview and were ready for the real fun. Patrolling through the woods, in squad and platoon sized echelons, was about to challenge us mentally and physically in a way that made the physical events of RAP Week seem trivial at most.

CHAPTER 3
DARBY PHASE
SEPTEMBER 2011
DEEP IN THE WOODS OF FORT BENNING

Camp Darby in September of 2011 was a grim place. What I can remember was arriving at a clearing from the woods and seeing tin sheet quonset huts lined up for each company where the students formed up in front of. An admin building was located near a parking lot for cadre vehicles. Besides this building, the only hard structure I recall was a latrine with open bay showers that looked uninviting in all aspects.

The students ate and slept around open bays that resembled three-sided plywood horse stalls. A squad occupied each bay for their planning use. Each bay had four large chalk boards and some planning tables and benches. Besides that, nothing shielded us from the elements. A burn barrel was located at each planning bay for warmth but reserved only for winter classes. We had just entered Darby Phase where we would learn how to execute squad sized reconnaissance's, ambushes, and attacks for the next two weeks.

Darby Phase consisted of roughly ten days of patrolling with one day halfway for refit and sustained airborne training for the airborne qualified students. Each day the squads were assigned either a recon or an ambush mission and their schedules revolved around planning, rehearsing, movement, and execution of the mission.

Three students at a time were graded, the Squad Leader, the Alpha Team Leader, and the Bravo Team Leader. The RIs broke the day up into two iterations of graded leadership for the planning phase and execution phase so a total of six students per squad could be evaluated. A "GO" from the RI meant you had performed to their standard in your position and could now advance to the next phase in the mountains of North Georgia near Dahlonega. A "NO-GO" meant you needed additional instruction and retraining and thus, were not ready to advance to the next phase of Ranger School.

This is known as "recycling" and added an additional three weeks onto the course. Students typically received two "looks" from the RIs and some even three if they needed their "GO". To graduate within 61 days, a student needed at least one "GO" per phase.

Throughout Darby Phase, we were incessantly "smoked" by the RI's during every briefing, training event, or class of instruction. Typically, we were given physical punishments while the class was formed into companies or platoons and receiving instructions from the RIs. It was to be expected. It wasn't anything personal from the RIs, they just wanted to see if you could endure being harassed and stressed out. So far, nobody in my squad had quit.

A typical day of patrolling started early at 0430 when the squad belted out the Ranger Creed. Every student was expected to know it by heart. Our cadre at IBOLC had us rehearse it incessantly until we got it right. To mess up the Ranger Creed was unacceptable to the cadre, Ranger School graduates, and the Rangers in the Regiment. To those unfamiliar, the creed is as follows:

Recognizing that I volunteered as a Ranger, fully knowing the hazards of my chosen profession, I will always endeavor to uphold the prestige, honor, and high esprit de corps of the Rangers.

Acknowledging the fact that a Ranger is a more elite Soldier who arrives at the cutting edge of battle by land, sea, or air, I accept the fact that as a Ranger my country expects me to move further, faster, and fight harder than any other Soldier.

Never shall I fail my comrades. I will always keep myself mentally alert, physically strong, and morally straight, and I will shoulder more than my share of the task whatever it may be, one hundred percent and then some.

Gallantly will I show the world that I am a specially selected and well-trained Soldier. My courtesy to superior officers, neatness of dress, and care of equipment shall set the example for others to follow.

Energetically will I meet the enemies of my country. I shall defeat them on the field of battle for I am better trained and will fight with all my might. Surrender is not a Ranger word. I will never leave a fallen comrade to fall into the hands of the enemy and under no circumstances will I ever embarrass my country.

Readily will I display the intestinal fortitude required to fight on to the Ranger objective and complete the mission though I be the lone survivor.

Rangers lead the way!

After this ritual, we then spent the next 30 minutes shaving, brushing our teeth, packing up our poncho liner (woobie) and preparing our sensitive items and platoon equipment for inspection. At 0500 hours, an RI who was assigned to monitor our patrol that day would approach the squad and announce by roster number who the Squad Leader and Team Leaders were. Then the squad received its Warning Order (WARNO) from the RI and began planning that day's mission in accordance with the Army Troop Leading Procedures and our Ranger Handbooks.

Throughout the morning, there was no eating or relaxing. Any students caught eating anything were automatically given a Serious Observation Report (SOR) which inevitably led to being recycled. Instead, the squad had to focus on writing the Operations Order (OPORD) and preparing the equipment. The Squad Leader made the plan, and the Alpha Team Leader planned the route to the objective and supervised construction of the terrain model. The Bravo Team Leader ensured the squad's weapons and equipment were in good working order and had all the ammunition and supplies necessary to complete the mission. The remaining seven members of the squad helped write the portions of the OPORD on the chalk boards in their planning bay, rehearse the key aspects of the mission with the Squad Leader, or executed tasks needed of them while leadership was busy.

In essence, the only way the squad would accomplish its assigned mission was if everybody worked together and made the maximum effort to ensure the success of the mission. By getting your buddies their "GO", they would help you get yours was the expectation.

I was assigned two positions during Darby Phase. The first, as Squad Leader, did not go as planned. Ripe with pride from earning the Distinguished Honor Graduate at IBOLC, I believed I had the knowledge to succeed in a squad sized mission. I greatly overestimated my skills and would pay greatly for it. I was assigned as Squad Leader in the second iteration in the day's leadership, for actions on the objective. After most of the day's planning and movement to the objective, the RI changed leadership and made me Squad Leader for an ambush mission. Although only 200 meters from the objective, things went downhill fast.

Soon after reaching what was the Objective Rally Point (ORP), the RI called out three roster numbers. For the Squad Leader, the number 257 was called. Finally, my moment had come. I knew what the mission was and felt confident in my ability to execute. Lacking camouflage face paint, I rubbed soil into my face to darken myself as the sun went down. Looking revolted, the RI said, "Ranger! You'll get a disease by doing that!"

I refocused from myself onto conducting the leader's reconnaissance of the objective site. Leaving a five-point contingency plan or GOTWA, I briefed the remaining members of the squad and took a select few out to finalize the plan. To my horror, the objective site was an open area in the woods, lacking any undergrowth for concealment or large trees for cover. I had to improvise, and fast.

Returning to the rest of the squad, I set in the left and right-side security in accordance with the plan. Next, I set in the M-240 Bravo Medium Machine Gun team behind a fallen tree. Finally, about five personnel were left for the assault force. There was no cover between us and the objective. We would have to move fast to destroy any enemy elements crossing our paths. And we waited. In the sunset of a long day, we laid in the prone with our weapons ready.

As the sun set, a single High Mobility Multi Wheeled Vehicle (HMMWV) approached on a gravel road. Once it was in the center of our kill zone, I signaled for the attack to begin. A training claymore mine was clacked and immediately followed by the scream of the machine gun firing. After thirty seconds, I called the weapon to cease fire. Sensing no movement from the enemy, I called for the assault team to advance. Immediately as they moved forward, a rifle began firing from the objective. As our team was now in front of the weapons team, I had no choice. "Move forward!" I called out. In the open, my assault team advanced online. Unfortunately, the RI was not having it.

As the assault element picked up and moved to the enemy position, the RI berated me and the squad with a torrent of obscenities. We were too slow. We had not suppressed the enemy long enough. We were disorganized. As the team moved closer, the man in the black t shirt with the tab assessed casualties from my team. First one. Then two. Three students who made it to the objective, were busy trying to clear bodies and the enemy's vehicle. The objective needed at least six to secure, and I only had three. After ten minutes, the RI had enough and called the codeword to end the mission, "ENDEX" or end exercise. The mission was over, but my torment had just begun.

Calling the squad in on the objective, I truly believed it had not gone that bad. After the RI had accountability of our weapons and equipment, he tore into all of us. Every aspect of the mission was ripped apart. From the recon to the assault, I was made to feel as if I failed my team. I felt ashamed and went internal. I completely lacked leadership as the RI told me that we now needed to move back to Camp Darby on foot in the darkness. My Alpha Team Leader, a regular Army Staff Sergeant, took over the route planning and successfully navigated the squad back that evening over a long trip through the woods. I learned a valuable lesson in leadership that night and resolved to never falter under pressure again.

The next day was an airborne operation, my first since jump school the previous summer while I was still a Cadet. A refresher jump, the purpose was to ensure all airborne qualified students were current prior to the tactical operations of the subsequent phases. We were allotted a decent night's sleep prior to the operation, of which I was grateful.

On a sunny day over Fryer Drop Zone on Fort Benning, a C-17 Globemaster would drop over 100 students from 1200 feet above ground level. After hooking up my static line, I remember looking over my left shoulder to see JT standing with the other half of the chalk about to exit the door opposite to me. I caught his eyes, gave a nod, and he responded likewise. I remember how nice the day was while I floated down to Earth, no longer a five-jump chump. I landed as smoothly as one could expect on a nice sandy pillow. My only lapse that day was when I forgot to remove the air items from the parachute for turn in at the rendezvous point. Although I got the look from hell from him, he checked off my name and dismissed me to my formation near a bus. I didn't hear anything about it ever again.

My next evaluation occurred five days later. I was assigned the morning Alpha Team Leader in charge of planning. As luck would have it, our Squad Leader was none other than J.T. Hess. Pouring myself into planning, I endeavored to help facilitate Hess's mission as much as possible. The terrain model under my charge was made into a work that

General Patton would have envied, complete with little twigs and leaves to simulate trees of the forest. It looked cheesy but damn was I not about to give this my all. When we finally stepped off after rehearsals, the RI switched out the leadership, thus ending my rotation. Though I felt confident in my evaluation, I would have to wait another week before I knew if I had my "GO."

Prior to stepping off on a patrol, we sometimes received a hot meal that had been transported from the DFAC at Camp Rogers. We were given a single plate and served our peers who in turn served us. It was never enough. Usually, we had five minutes to consume everything. There was no time for conversation. One time, after our five minutes had elapsed, the RI told the entire platoon, "Rangers, you have sixty seconds to finish this chow. Begin." There was a mad rush to the serving line where the remaining containers of food remained. There wasn't enough for everyone to get a portion. It was complete chaos. Students were grabbing containers for their squads and hoarding them. Some were dipping their nasty hands into mashed potatoes. Others were desperately trying to chew as fast as they could. Some choked. Some threw up. Others cursed that they hadn't received anything. Those that grabbed a protein bar struggled to consume it.

I went by another approach. The gravy had not been raided by the time I worked my way through the hoard to the serving line. Grabbing a Styrofoam cup, I dipped it into the vat and began drinking gravy like Coors Light at a frat party. It was delicious and easy to put down in bulk. It also was packed with calories and fat, which I desperately needed. After two cups, I heard the "Time Rangers!" from the RI.

Everyone immediately vacated the line and returned to the formation pit. Nobody carried anything with them, the RIs scanning for any sneaky students who thought they could pocket a piece of bread or chicken. Then we were smoked for leaving the serving line a complete mess. I must have thrown up brown gravy about three times between the pushups and calisthenics.

After a patrol, we sat on the rocks in front of the Bravo Company quonset hut. We returned shortly after midnight every morning and cleaned our weapons and refitted our kit. On a particular Saturday night, one of the RIs, the Company Commander of all people, wanted to play a game with us. Observing us cleaning, he asked the whole formation in a friendly tone what day of the week it was. Confused by the question, we responded with "Saturday, sir!" The Captain then asked, "And what do we do on a Saturday night, Rangers?" He was so cool in the way he asked, like he was just one of the guys. One of the young students from the Ranger Regiment proclaimed loudly "We get fucked up sir!"

"That's right Rangers!" the Captain announced as his calm and collected attitude suddenly morphed into a crazed grin. "And now I'm going to fuck you up." We had fallen for it and paid for it dearly. He smoked us for what seemed like an eternity. Once satisfied with his joke, he disappeared again into the hut. That was what my Saturday nights were like in Ranger School.

Finally, we completed our last patrol for the phase. Our reward was more hours of cleaning weapons and equipment on the rocks in front of the company quonset hut. After additional smoking's for the most minuscule of deficiencies, we finally were allowed a break in the form of a DOGEX, a traditional cookout per phase where Ranger students can purchase hotdogs, candy, or soda pop for $1.00 apiece. It was most welcome after almost three weeks of intense training and little sleep or food.

I found out I had received my "GO" after a thirty second counseling with an RI and was thus destined to move on to Mountain Phase at Camp Merrill in Dahlonega, Georgia. Our squad lost three members to recycling that day. The rest of us carried ourselves high, with a swagger of sorts as we boarded the buses for the mountains of northern Georgia.

We were permitted to eat our MRE on the bus. The previous evening, all Rangers received their mail and any care packages. For thirty minutes, we gorged ourselves on home baked goods and junk food. We ate so much and so quickly that many of us became sick later that evening. After a taste of real food, the MRE sat in my assault pack for the duration of the bus ride. It wasn't until a rumor floated around that they would take the unopened MREs away later did I finally consume its contents.

As the buses moved north, many of us slept. Overcome with the excitement of passing Darby Phase and going into the mountains in that early October, I stayed wide awake. I wrote Michelle a letter explaining how I was, even though I had called her the previous evening to inform her I was moving on. I missed her dearly and could not wait to see her at graduation on November 10th.

CHAPTER 4
MOUNTAIN PHASE
OCTOBER 2011
DAHLONEGA GEORGIA

At dusk, the buses pulled into Camp Merrill, home of the 5th Ranger Training Battalion and located deep in the Appalachian Mountains. The base itself looked vastly different than Camp Rogers and was spread out more. One thing was for certain, it was anything but flat. Pine trees lined the perimeter, and the ridgeline obscured any view of the horizon. We arrived at two huge two-story buildings that would serve as our barracks when we were not patrolling. For the next three weeks, this would be our home.

The first five days of Mountain Phase were designed to introduce students to techniques in mountaineering. Mornings consisted of wake-up around 0430, hygiene, marching to the DFAC, saying the Ranger Creed, breakfast, and then a quick rush to the barracks to grab the packing list for the day's training. All of this was conducted on a section of Camp Merrill called the "uppers." The actual training was at another area which we moved to on foot down a steep trail to the bottom of a valley. This area is known as the "lowers."

We learned the prerequisite knots for basic mountaineering, tying

"Swiss Seats," commands during rappelling, and how to rappel with combat equipment. We lowered casualties, constructed apparatuses to slide Soldiers and equipment quickly down steep inclines, and built one rope bridges to cross creeks. Although the days were cold and we shivered incessantly, the RIs attitude towards us was quite different than at Camp Darby. In fact, we were smoked less than twice a day and I venture to say that even then the smoking was light.

One of the best aspects of Camp Merrill was their dining facility (DFAC), which served some quality breakfast and dinner when compared to the open-air Army chow of Camp Darby. Then there were the fabled blueberry pancakes of Camp Merrill that Ranger students speak of in high regard. I can confirm, they are amazing and hit the spot on mornings prior to stepping off to train. We even had slightly more time to consume our meals in the DFAC, unlike in the first phase where RIs are screaming at us to chew, chew, swallow, and repeat.

The final two days of our introduction to mountaineering occurred at Mount Yonah, Georgia. The entire class was bussed to Mount Yonah, roughly an hour away, thus more sleep. After the students hiked to the summit, we conducted practical exercises in rock climbing, rappelling, and rescue techniques off the cliff face. Our rucks were heavy with all of our equipment, well over fifty pounds, and we hadn't even been issued our platoon equipment for patrolling. Still though, it was some of the best training I have ever had and the weather up to that point was gorgeous, despite the low temperatures.

I attended my first church service on top of Mount Yonah since commissioning where the Ranger Chaplain distributed communion in the form of a piece of bread and some grape juice. After the last three weeks of hell, it was no small wonder the service was packed with denominations of every sort who yearned for spiritual relief.

To accommodate those recycled from Camp Darby and the Mountain Phase recycles, the platoons were reformed, and personnel moved to different companies. Unfortunately, only one member of my squad was reassigned, Ranger Hess. He was moved to another platoon adjacent to us. We received three new Ranger students into our gang.

On the day prior to executing our first patrol, now as a platoon instead of a squad, I witnessed a student quitting for the first time. To everyone's shock, he walked away from us and abruptly told the RI he was done. Surprisingly, the RI showed him dignity and respect and the student waited to be picked up and moved back to Fort Benning. I couldn't believe it. He had made it this far already. He wouldn't be the last to quit the platoon by the end of our trials and tribulations in the steep and dense mountains of northern Georgia.

Patrols were broken up into two five-day iterations and separated by one day for refitting equipment. However, unlike at Camp Darby the students stayed out in the woods instead of returning to base in the evenings. The patrol base, not the planning bay, became our Motel 6 for eight out of ten nights. Along with reconnaissance's and ambushes, students were now expected to plan raids that attacked enemy positions and conducted a hasty withdrawal off the objective. Team Leaders were no longer evaluated. Only the Platoon Leader, Platoon Sergeant, and the Squad Leaders were graded. Additional non graded appointments were made in the Forward Observer, Medic, and Radio Telephone Operator (RTO) roles.

The day of our first patrol was hell on Earth. Our mission was to move dismounted from Camp Merrill to a location on top of a large feature known as Hawk Mountain. Our rucks weighed close to 80 pounds when we stepped off, and to top it off, a steady rain began to pour. The infiltration was slow, and the winds picked up. Temperatures plummeted as we climbed up the incredibly steep terrain. There was no trail and we prayed to God that we didn't misstep, as the tumble under all of our gear would have surely caused severe injuries, or worse.

By the time we made it to the summit and thus, the Objective Rally Point (ORP) prior to the Leader's Recon, many went internal. The RI who remained behind at the ORP to observe the platoon told everyone they could stand up to keep warm but were absolutely forbidden from eating or leaving the ORP. He then promptly sat on his camping stool he hauled with him, put his rain jacket hood over his head, and bent over as if he was sleeping. Even the lot of the instructors can be miserable apparently.

For the next five days, with temperatures in the lows of the mid-thirties and highs of upper fifties, we trekked under water-soaked rucks throughout the lower Appalachian Mountains. Each day started at 0430 hours when the platoon came together in the center of the patrol base to layout all its weapons and equipment for inspection.

Like a rooster announcing the arrival of the sun, we sang the Ranger Creed as our welcome to the RIs coming to evaluate us for that day. From 0600 until 1200, we planned our day's mission: a raid, ambush, or recon. Those participating in the planning process gathered in the center of the patrol base to build the terrain model and develop the plan based off maps that had entire regions missing or faded to obscurity. The remainder of the platoon pulled security on the perimeter of the patrol base, hoping desperately for the sun to warm their tired, hungry, wet, and cold bodies.

As with Darby Phase, consuming any substance besides water was strictly forbidden with the penalty of being recycled. However, the cold weather now brought another temptation into our midst. Although on the packing list, Gortex underwear and undershirts were not authorized for wear without approval from the RIs. The reason was simple enough, students often wore them at the wrong temperatures. The result was that the Gortex caused the wearer to overheat and could possibly become a heat exhaustion or stroke victim. As a result, we were not authorized to wear Gortex after 0500 hours when we rose from our sleeping bags under our ponchos in the rain. However, as with food, students were often caught by the RI's wearing these garments. They were immediately given an SOR which meant the dreaded recycle status.

After 1200 hours, we stepped off for our mission. Often between

8-12 kilometers away, we walked to our ORP which was only one kilometer from our destination. While some companies had helicopters and trucks to move them, I can recall getting no transportation for the first five days. As a result, our feet began to suffer from our waterlogged boots. Although we changed socks multiple times a day, our feet began to develop large blisters and a case of trench foot was not uncommon. Foot powder turned quickly to a yucky paste in the rain if a student forgot to put up his poncho to shield him from the elements.

After accomplishing our mission at around six pm in the evening, we conducted another movement of roughly five kilometers to our patrol base site. From there, we conducted reconnaissance, set in security, and conducted priorities of work. On average, we finally fell asleep in one-to-two-hour shifts at 0200 hours. With the rain, cold, lack of food, and little sleep, we were quickly losing our ability to function as a cohesive team.

The first three days of the FTX were plagued with issues for my platoon. Due to the rain and cold, many students broke down and went internal. Quite a few were caught eating articles of chow from the MREs, with one even caught with a small bottle of Tabasco Sauce. If that wasn't enough, we had a string of SORs every morning from the patrol bases. The M-249 Light Machine Guns, a staple weapon for the infantry squad and used by the Automatic Riflemen, seemed cursed by a select few ill-fated souls.

When the machine guns were switched out in the mornings, some of the Rangers had halfcocked the weapons with the ammunition in the feed tray. Whenever a weapon was jolted or moved briskly, a few rounds were expended through the barrel of the weapon in a loud string of "BAM...BAM...BAM!" Known as a Negligent Discharge (ND) when the weapon fires without a purpose, this is one of the most egregious acts a Soldier can conduct due to the possibility of injury or death. The students who had this happen to them were immediately isolated and returned to base.

After a pattern of at least one ND a day, we had an entire fire team of our platoon missing and everybody feared the position of Automatic Rifleman. When my turn came, I cocked the lever fully to the rear, placed the weapon on safe, and allowed the blank rounds to stay within their pouch and out of the feed tray. It wasn't until a stern rebuke from an RI later that day that I placed the rounds in the weapon to ensure I could engage the enemy whenever I needed to. For the entire time, I held the charging handle to the rear, fearing a malfunction in the weapon that would send the bolt forward and fire a few rounds.

As if the fates of the students who had NDs weren't bad enough, the Command Sergeants Major of the 5th Ranger Training Battalion ordered the soon-to-be recycled students back out into the field to continue patrolling. Even though they were being recycled, they could still train with us was the rationale, instead of being warm and dry. They could either quit or train. I do believe the van returning the students was lighter and our platoon still lacked additional manpower despite the offer.

My evaluation came in the middle of the first five-day field training exercise (FTX). I was assigned as the Platoon Sergeant for a raid after entering the ORP. The RI even hinted, "I hope you know what a Mass Casualty (MASCAL) event is." I set to work as the Platoon Leader initiated his leader's recon of the objective. Going to every position with a Squad Leader, I personally oversaw the repositioning of weapons and claymores, the reapplication of camouflage paint and foliage on helmets, and the lubrication of all weapons systems. We moved ruck sacks into the center of the ORP after the reconnaissance element reported that the objective location was confirmed.

All of the platoon specialty items of equipment were prepared and distributed for the mission. I felt confident in my team and pulled out all the stops to ensure mission success through our readiness.

The Platoon Leader returned to report his findings and immediately set about moving the squads into place. Imagine a deadly game of chess where the squads are positioned for the checkmate. Security was set in first. Followed by the support by fire element consisting of the machine gun teams. Finally, the two squads that assaulted the objective were positioned at their attack points. The Platoon Leader with the assault element, I moved myself to the support by fire position with the assigned medic.

At a designated signal, the support by fire element opened fire on the enemy locations. Known, likely, and possible positions were targeted and engaged with a hail of blank 7.72 rounds from the two M-240B Machine Guns. The Weapons Squad Leader (WSL) was an experienced Ranger Regiment private, and it was obvious he knew his trade.

The guns sang beautifully as they took turns each firing volleys onto the objective and picking up the rate of fire as guns changed smoking barrels. I saw a green flare fill the overcast sky over the assault position, the signal for the gun teams to shift fire to open a corridor for the Ranger students to hit the objective. The WSL acknowledged the flare, had his gun teams shift to another target, and then fired a white flare into the sky so that the Platoon Leader knew for certain that we had seen his flare and shifted.

Two squads bounded forward in teams until they reached the objective. By the time the first team reached the wood line, another flare dazzled the sky, the signal to cease firing on the objective. Immediately all firing from our gun teams stopped. The Ranger Private charged with two-gun teams, fired his last flare. His mission was accomplished. The two squads on the objective quickly cleared one little hut and a couple vehicles. Search teams scoured the opposing force (OPFOR) bodies and the vicinity. There is an old wives' tale that anything food related that is found on the objective is fair game. The OPFOR had three giant bags of chips that were quickly secured for devourment when we were permitted to eat after priorities of work later that evening in the patrol base.

Suddenly, a loud "BANG!" shook the area. The RI had thrown a simulator which sounded like a fragmentation grenade cooking off. Looking at the Platoon Leader, the RI explained "Ranger, those chips were booby trapped and your search team has been severely wounded. They cannot walk or use their arms. Evacuate them down the road for a MEDEVAC pickup. Oh, and those chips are destroyed. Leave them."

The Platoon Leader quickly called me up on his radio. "We need the medic down here fast. I have four wounded." We moved at a brisk pace to the site of the explosion and immediately went to work. After the medic assessed all casualties, I spun up a 9-Line MEDEVAC report to the RI and received the new coordinates for the pickup site two kilometers away. Our aid and litter teams quickly put the four students onto portable litters (SKEDCO) and had them secured for movement.

"Rangers, move by the road to the pickup site." What luck we had. Not through the woods but on a road. This RI must be in a good mood I reckoned. The Platoon Leader called in the support by fire element. Within three minutes, the squad was ready to move, their leader had already taken the 240's off their tripods and readied them for the exfil. Then the security teams were pulled in with one team leading from the front and another from the rear. After five minutes of quick action by the Platoon Leader, the platoon was lined up and ready to go.

"Hey man, can you lead the element from the front since you planned the route?" the Platoon Leader asked me. "No problem, brother," I replied and pulled myself to the front of the element. I counted off all members of the platoon on my way to the front and we moved at a brisk pace. Since our rucks were still at the ORP, we felt much lighter than usual. After twenty minutes of carrying the SKEDCOs, we were ordered to halt. "ENDEX!"

The RI called out for us to lay out all weapons and equipment and get another count. The ORP was less than 500 meters away from our position and, moving by squads, we retrieved our rucks and the two Rangers guarding them.

After twenty minutes of chaotic movements throughout the Appalachian wilderness, over forty Ranger students stood in two long lines that surrounded the platoon's weapons and equipment on the ponchos. A steady rain had begun falling. Satisfied with the count, the RI ordered us to pull out an MRE and listen. Forty students dug into their rucks and pulled out their most treasured possession.

"I didn't say eat it, Ranger!" he barked. For ten minutes, we sat with our dinner in front of us as the RI went over the mission and what he thought was satisfactory and what needed improvement. My heart soared when he exclaimed that the actions in the ORP were the best he had seen. He said the MASCAL, while at times chaotic, moved out the casualties for pickup with urgency. I felt confident in this evaluation. Maybe I had my key to Swamp Phase.

"Put your MREs away!" With drooping faces, we obeyed his command and packed up our much longed for food again. That was a huge tease I thought. A hail of roster numbers was called to replace the platoon leadership and designated positions. I turned over my map and platoon equipment to the new Platoon Sergeant.

Our next mission was to move to the patrol base site about four kilometers away. The rain had now begun pouring and the evening had caused the temperatures to drop into the low forties.

"Tonight, will not be fun," I said to myself. After ten minutes of tying down and distributing weapons and equipment, the Platoon Leader briefed the Squad Leaders on the site and route. The route just happened to parallel the Appalachian Trail for most of the way. "We will stay five meters off the trail and use it as a handrail," our fearless leader exclaimed. The RI, not enjoying the weather any more than us retorted, "The hell with that Ranger, take the damn trail." And so, for the first time in my life, under the most unusual of circumstances, I hiked the Appalachian Trail. Or at least a few kilometers of it.

I was never so tired that evening as I took my first steps on a route I had always longed to trek. The waterlogged rucks now weighed closer to 90 pounds and every muscle and joint in our tired starving bodies screamed with any movement. Nonetheless, we weren't breaking brush.

After two hours of hiking in single file, we departed the trail and set up the security halt for the patrol base recon.

After three more hours of recons, security, and priorities of work, we were finally permitted to get some sleep. I was first on guard and stood up on my feet to avoid passing out. My night vision device (NVD) completely fogged out from the weather, I just stood there in the rain staring into a green mass in front of me. It had been a good day overall I thought.

While shivering to the bone and sore like never before, I felt confident in myself. I wanted to take a platoon in the 82nd Airborne Division. My Ranger Tab was the validation needed for that blessing to lead Soldiers I had been told time and time again by my instructors. "I don't have time to recycle," I told myself. I needed to be out of here with a black and gold tab in November. Finally, I was relieved and passed out in my sleeping bag. We had one more day of patrols prior to our refit day. "I got this," I told myself as my eyelids met for the two hours of intermittent rest.

One night, after a strenuous day of patrolling, in a torrential downpour, we were going through priorities of work in the patrol base. The Platoon Leader, not thinking clearly, allotted only ten minutes for the entire platoon to eat. Because we had to maintain security at all times, only one third of the platoon could eat at any given interval. Each squad broke the time up into three-minute portions with the Squad Leader getting a minute to consume something. Unfortunately, the first group did not eat quick enough, so the second group received even less time. By the time the third group was about to eat, the RI told the platoon to put all food away. I was in the third group.

Later that night, after pulling my shift of security, I reached for the opened MRE in my sleeping bag and pulled something out. I was starving. I pulled out what I thought was jalapeno cheese spread and opened it up while huddled in my sleeping bag. I was so tired that I didn't even notice that it was not cheese spread but a vanilla protein shake powder.

I didn't even realize I was also opening it above my head. Inside of that sleeping bag in the early hours of a rainy morning, confused and saturated, I opened that packet. An explosion of white powder covered my face, neck, and the interior of my bag. And because I was completely wet, the powder immediately turned into a nasty creamy vanilla frosting on my face.

Mentally defeated, I stood up and allowed the sleeping bag to fall off of me to my feet. Not more than three feet from me was an RI. He shined a red flashlight in my face and looked me up and down. He said something along the lines of "How do you have any energy for that, Ranger?" He shook his head and promptly walked away, leaving me alone as the pouring rain cleansed the white vanilla mess off of my face and upper torso. I'm pretty sure he knew what I was up to, but the packet was still in my sleeping bag, about the only saving grace I had that night. To this day, I wonder how many of his buddies were told about the Ranger Student who frosted himself on a cold and wet October morning.

The 24 hours of refit was a day of plenty for the class. After cleaning weapons and equipment until they met the high standards of the black shirts, we were permitted some rare luxuries. A DOGEX was conducted that day and for the always low price of $1.00 per item, we gorged ourselves on hot dogs, chips, cookies, candy, and soda. This treat sweetened the day as we had hot chow for breakfast and dinner in the DFAC and the complementary lunch MRE which we still devoured. If that wasn't enough, mail and care packages were distributed that evening, and we were allocated a whole hour to consume the delicious contents of our packages.

My father had sent me a package containing a letter and the weather report for Dahlonega. The next five-day FTX was all sunny skies and warmer temperatures. Since we couldn't access the radio or internet, the weather was unknown to everybody but the RIs. I read aloud to the entire platoon of the future forecast for the next week. "Three cheers for

Ranger Pitcher!" erupted and a chorus of "Hip hip, hooray!" filled the air. Finally, after the last five days and nights of utter misery, our entire class was getting a win.

The second FTX passed without incident. We were trucked by Light Military Tactical Vehicles (LMTV) on our first mission on a beautiful fall sunny day throughout Northern Georgia. It was simply gorgeous. Despite the aches and the incredibly heavy burden on my back, the weather and terrain made me feel invincible. Growing up in the woods, I savored this moment as one of my dreams was to hike the Appalachian Trail in its entirety. Now I was reconnoitering it for future adventures with my wife and children.

My second graded patrol was on the third day of the second FTX. Immediately after a raid, the RI assigned me as a Platoon Sergeant for a patrol base mission. "Look Ranger, this guy still needs his GO. Help him get it," I was informed. "Roger Sergeant!" I exclaimed. It wasn't a direct confirmation, but I had indeed passed my patrol a few days earlier.

After two hours of walking in pitch black wilderness, a miracle happened. "Ranger, line up all rucks in two columns. Each Ranger sleeps at his ruck and stays in the lines. Starting from the ends, assign a guard for watch and place everybody to sleep." An administrative patrol base! What luck! The entire mood of the platoon turned north. No priorities of work, no security, no reconnaissance's. Just eat your MRE and pass out until awakened for duty.

Finally, after our last ENDEX, we were trucked back to Camp Merrill. Still not knowing our fates, we all immediately got to work cleaning barracks, weapons, and turning in equipment. In two days, the students who passed Mountain Phase would move onto Swamp Phase at Camp Rudder, Florida. We were given our mail and care packages. To raise more money, the 5th Ranger Training Battalion even authorized a second DOGEX which we enthusiastically supported through a cash infuse.

When the day of reckoning occurred, we lost two members of our squad to being recycled. I was not among them. But the training wasn't over yet. Our next mission was briefed to the class that the airborne qualified students would be bussed to an airfield and dropped into Camp Rudder by C-130 on a tactical airborne operation. The remainder would bus from the mountains to northern Florida.

CHAPTER 5
SWAMP PHASE
OCTOBER TO NOVEMBER 2011
EGLIN AIR FORCE BASE, FLORIDA

Knees in the breeze, I exited a C-130 Hercules at 1200 feet above ground level and had the softest landing ever on a sandy drop zone. The contrasts in the terrain around me on Eglin Air Force Base when compared to Camp Merrill were vast. Completely flat, pine trees and sandy soil replaced the beautiful Appalachian scenery I just escaped. Walking to the assembly area, I observed an airfield, fire station, and a road of white buildings that made up Camp Rudder. A large four-story barracks was now our mansion for the next two weeks. Hess had also made it to Florida I found out soon after. Two boys from the Bluegrass State still in the running for the coveted black and gold tab and still on track to graduate in three weeks' time.

The accommodations at Camp Rudder were the best yet in the training phase of Ranger School. The barracks were relatively new and the showers, although open, did not give the user a potential case of tetanus when operated. Even better was the DFAC on the camp. If the blueberry pancakes of Camp Merrill were amazing, the french toast was simply outstanding. The battered bread was heavily dusted with cinnamon and sugar with a thin coat of butter. I personally witnessed students trading their entire breakfast of eggs, grits, bacon, and everything else just for those two pieces. They were that awesome and I still haven't had a batch of french toast that competes with this culinary masterpiece from Camp Rudder.

A zoo of sorts also exists on the camp. As odd as this sounds, the reptilian wildlife that inhabits the panhandle swamplands is on display by the Ranger Instructors of Charlie Company, 6th Ranger Training Battalion. Venomous snakes such as the Eastern Diamondback Rattlesnake, Water Moccasin, and Copperhead reside in their terrariums. Full sized alligators also have an enclosure complete with outdoor basking pools. It was quite the setup and served as an educational opportunity for the students and a means to train the instructors in what to look for on the trail while guiding the platoons through the swamps.

The legendary animal activist from Australia, Steve Irwin, had even visited the facility where an episode of his show, Crocodile Hunter was filmed. We were given an awesome block of instruction on the reptiles of the area. At one point, an Eastern Diamondback Rattlesnake was mere inches from my face as the handler brought the serpent around the students. I surely hoped I wouldn't encounter one of these bad boys on the trail down here.

Swamp Phase back in 2011 lasted roughly two weeks. After arriving at the camp, students received four days of practical exercises in swamp crossing, patrolling in the area, and utilizing zodiac watercraft to infiltrate to the objective area. Just like in Darby and Mountain Phase, each platoon was given detailed instructions for how to execute a specific mission and the best tactics, techniques, and procedures (TTPs) to employ. Only this time, the RIs in Florida were more hands off. The expectation was that by now, we should know the basics enough to execute a mission.

The RIs now refined our platoon's TTPs before ratcheting up the difficulty of the operations we were to execute during our FTX. There was no yelling. There was no smoking. It was as if we were seen as equals in their eyes. Grazing, or eating portions of your MRE during a mission, was even authorized, to an extent.

The last ten days of Florida Phase were a nonstop FTX in the northern Florida swamps. No refit days, and no barracks or hot chow for ten whole days where our missions increased in length of movement and difficulty in execution. We were to infiltrate the enemy area of operations by airborne assault and execute raids, ambushes, and reconnaissance's throughout the FTX.

The final mission was to be a raid on an enemy island where students utilized zodiac watercraft to infiltrate to the objective area. Each student was given ten MREs to pack into their ruck for the first five days of the FTX. A resupply would be conducted halfway through. If you ate all of your food in the first two or three days, you had none until you were resupplied or the FTX was completed. Each student also packed a small bag of dry clothes for the resupply that was secured in a container and would later be trucked in after a swamp movement.

On the morning of our expected jump, I weighed my ruck. 110 lbs. of equipment, ammunition, food, water, and clothing. My ruck looked like a severely bloated tick. The plastic buckles strained as I pulled with all my strength to secure my mobile home for the next ten days. Luckily for the other Paratroopers in our midst, the weight of the rucks was too heavy to jump with. Instead, we jumped with lighter assault packs that weighed roughly 30 pounds, thus saving our knees, and affording us the best opportunity to have a safe exit from the aircraft. Prior to loading the aircraft, the 6th Ranger Training Battalion Chaplain led us in prayer. He then rigged up to jump with us. Now that's the kind of preacher man I want leading my sermons, I reckoned.

On the evening of our first day of the FTX, we boarded C-130's for the operation. The flight itself was short. Elizabeth Drop Zone was located a few kilometers south of the camp. The weather was warm and sunny, and the light breeze made for a perfect jump. When the red light turned on, the black shirted Jumpmaster sounded off with, "Get ready!"

Our attention to him, we waited for the series of commands to prepare us to jump. After being ordered to stand up, hookup, and check static lines and equipment, we were ready to assault the drop zone. On the green light, we exited briskly under the watchful eye of the Jumpmaster. After an initial shock, my parachute opened perfectly, and I floated down to Earth. Unfortunately, as fate would have it, my luck almost ran out on the very evening of my final FTX in Ranger School.

As my parachute came down towards the drop zone, I lowered my equipment at about 200 feet from ground level. It was dusk and a light breeze was making me float to my left. I pulled on my risers and hoped for a soft landing, as my previous two had been in the course. But that wasn't the case at all. Instead of hoping, I should have been praying.

I hit the ground like a meteorite and felt it in every bone of my body. Although my chin was tucked in, the force of the landing and the density of the point of impact forced my head back and into the Earth. After going motionless, with my chute coming to a rest like a shroud over me, all went black.

I came to soon afterwards with a large headache. My head was wet, and my body was battered. Despite the force of the landing, the adrenaline set in while I removed my air items and packed up my chute. I reached into my assault pack for my NVDs. As I attempted to mount them to my Advanced Combat Helmet (ACH), I saw that the mount had completely sheared off from me hitting that iron plate of a drop zone. I put my weapon into operation and followed a large group of Ranger students to the turn in point. The look from the RI said it all when I approached with my chute and weapons case.

"Holy shit, Ranger, you look like hell!" the RI informed me. "Roger, Sergeant," I bleated out. "No, dumbass, go see the medics over there," he countered, obviously concerned something was wrong. I moved to a Field Litter Ambulance (FLA) where two medics were stationed. Immediately the medic removed my helmet and began taking vitals. I explained I had hit the ground like a lead brick and that nothing seemed broken except my helmet's baseplate.

However, my body wasn't what they were concerned about. I was leaking yellowish clear fluid from both ears, a sure sign I ruptured my ear drums and had a concussion. I told them I just wanted some Naproxen for the pain and to continue training, but they advised against it. "You need to rest a couple of days," the senior medic said.

Knowing this would be tantamount to recycling, I politely declined. "Suit yourself, but drink plenty of water and get some rest tonight." I did not know it at that time, but nobody would be sleeping that first night.

I met up with the rest of the platoon who was forming up in the tree line. All the non-airborne personnel (NAPs) were unloading a large box truck that had all of our rucks. Finding my ruck within the giant mass of bloated ticks, I set about cross loading the equipment I had jumped with. I assumed my position on the perimeter as a Grenadier in my squad, my M320 Grenade Launcher at the ready under my M4. My head was killing me, and I couldn't mount my NVDs. With darkness quickly setting, I hoped with everything that this would be our patrol base or relatively near it.

But that was not the case, the Platoon Leader briefed the route. We were moving over 12 kilometers on sand trails that night. Known as the "walk to daylight," these gut-checks bypass patrol base operations in their entirety and force the platoon to cover long distances throughout the night. And the earth was not the rocky soil of Dahlonega or red clay of Columbus. No, it was sandy. And with rucks weighing over 100 pounds, walking on loose sand was miserable on our feet and legs in every aspect. If that didn't seem bad enough, I was blind in the dark and pretty sure I needed a CT scan for my head.

With one sinking and sliding step after another, we covered the distance. I was weaving in and out of reality as we pushed through the movement. I kept thinking of Michelle, and how much I wish she was here right now. She always knew how to comfort me. I had nothing in my ears to stop the flow of fluid and the headache persisted with intensity. "I can't fall back," I remember telling myself. "I'm so close to the finish line." I continued walking behind the student in front of me. The cat eyes and illumination tape on his ruck and helmet guided me forward. The only sounds made were footsteps and the occasional readjustment of a ruck on my back.

Twice on the route, the enemy force hit our platoon in pairs with harassing fire. Instinctively the entire platoon reacted to contact and took the enemy out. It was more a nuisance really because it slowed our progress and caused us to have to regroup each time and get accountability of equipment.

Luckily for us, after doing these kinds of missions for weeks, we had become very efficient at destroying small enemy outposts and consolidating afterwards. After getting accountability for each attack, we set off again on our journey deeper into the Florida woods. We did no water crossings or encountered any swamps that evening, which proved to be most fortunate. We had to cross multiple Linear Danger Areas (LDA) during that trip. Fire breaks or paths cleared in the forest to contain forest fires had to be secured on the near and far side and crossed tactically. After about four LDA crossings, the RIs, also tired, advised us to cross the remainder as if they weren't there. They were also on a timeline and had to get us into position.

Finally, we stopped for a security halt and were told to pull security. I could hear engine sounds from tactical vehicles nearby, a sure sign that you are near a logistics transfer point and thus, your destination. It was around 0400 in the morning. "Rangers, you have one hour of mandatory administrative sleep," our RI told us. I dropped my ruck against a tree and passed out on it. I didn't care if I woke up or not after that evening. When we were awakened, medics checked on the entire platoon. I was given more Naproxen and an RI brought me a new baseplate which I struggled to mount with my Leatherman.

The next few days passed in a blur. Each night brought little sleep and the movements remained long throughout the day. Unlike other companies, Bravo Company didn't have helicopter or vehicle transport that phase. We had no "Golden Walks," special events during an RIs last walk where the students are brought food to feast and GOs are supposably given to all leadership. We had no "Chow Birds," aerial infiltration or exfiltration movements where the crew chiefs feed pizzas or burgers to their passengers. Our only saving grace was that in this era of Ranger School, grazing on food was allowed in Florida Phase.

My first evaluation in Florida was as a Platoon Sergeant for morning planning and actions on for a raid on day four. I remember little except I forgot to check the range cards for the weapons squad. An RI casually pointed out to me that the barrels of the M-240 Bravos were pointed in a completely different cardinal direction than what was annotated on the range card. The OPORD and the raid turned out to be successful, but this oversight cost me a "GO" on my first patrol.

One memorable instance for that long FTX was during a zodiac insertion down the river. The 6th Ranger Training Battalion Chaplain accompanied my squad. In silence, as we paddled, he passed each one of us a small treasure. A single jellybean landed in my gloved hand. I placed the small morsel on my tongue and felt a strange high from the sugar as it rushed to my head. Although I am not a fan of jellybeans, that gesture of kindness stays with me to this day.

We did two swamp crossings at dusk prior to hitting our objectives in the evening. At two locations, known as Boiler and Weaver, I held my weapon over my head as I navigated waist to shoulder deep murky water. A rope secured to the other end of the swamp served as my guide. It was getting dark so we hurried as fast as we could. Bubbles ascended from the water everywhere. Water moccasins and rattlesnakes became my primary fear as I waded through this mess.

Luckily, the entire platoon passed through this terrain feature without incident. It was early November, and the water was cold. As soon as we reached the dry bank of the swamp, the whole platoon was ordered to administratively change into their dry set of clothes that they had previously set aside. That was the first dry and clean uniform I wore in almost five days.

On the night before our final mission, I received my second look. "257!" My roster number was called out after a short halt on a movement in the late afternoon. "PL!" The RI Captain who must have been the Company Commander informed me. My mission was to execute a raid on a small village of three buildings that consisted of shipping containers in the woods. I looked at my map. The objective wasn't even on it. A giant hole obscured my ability to see where I was going. "Sir, can, uh, you throw me a bone?" I asked. "What, Ranger?" I pointed out the map's handicap. "That sucks," he exclaimed.

Instead of berating me like I expected, an angel from mercy appeared in the form of another RI who showed me his map. The platoon was only one kilometer from the objective. Darkness was setting in fast on this early November day. I looked around at the platoon still in the short halt and advised my Platoon Sergeant and Squad Leaders that this was now the ORP. We couldn't go any closer without compromising the mission. I brought in the leader's recon and left a Five-Point Contingency Plan to the Platoon Sergeant. Due to the flatness of the terrain, we covered the one kilometer in no time. I pinpointed the objective from a tree line and observed other locations for the support by fire and security elements.

Returning to the ORP, we briefed the completed plan to the leaders and set about emplacing our chess pieces. Security first. Then support by fire. Then assault with myself leading the attack. Talking to the student serving as the Forward Observer, I called a notional artillery mission with the objective to soften up the OPFOR. Three artillery simulators whined loudly before a loud "BANG!" I had the guns open fire after the rounds impacted to suppress our objective. After a brief light show of green and white star clusters to signal the Weapons Squad Leader to shift and lift firing, we began clearing the objective.

All seemed to be going well. After five minutes, the attack was over with two casualties assessed. I called the Platoon Sergeant forward with the medic and had the aid and litter teams get to work. After a 9-Line MEDEVAC was submitted and the casualties prepared for movement, we were ready to exfil.

As our deliberate withdrawal was about to begin, I heard "ENDEX!" This was followed by a chorus of "ENDEX!" from the rest of the students. "Bring it in, Rangers," the RI Captain said. Looking at me, he ordered that all equipment and personnel was to be brought from the ORP and laid out for inspection. At last, it's over I believed. Hopefully my second look was satisfactory in the Company Commander's eyes, I reckoned.

But instead of announcing new roster numbers for the patrol base movement and occupation, I was briefed to move the platoon with the casualties down a sandy road for roughly four kilometers where LMTV transport would pick us up for movement to our patrol base. 'Roger Sir!" Lesson here is that it's not over until you're relieved.

Due to being day eight or nine of the FTX, the entire platoon was smoked. That movement became a nightmare for every student as we struggled to evacuate our comrades and carry their burdens. I moved up and down the formation, trying desperately to motivate the team as we slogged our way to our link up point.

Thankfully at this point my baseplate had been replaced and I could see in the dark. Witnessing some LMTVs up ahead, I ordered a short halt and sent a party to conduct a linkup. "Just get on the trucks, Rangers!" the Captain ordered. Obeying his command, we picked up and loaded expeditiously. Every single one of us to a man fell asleep on those trucks. I was awakened by the sound of a parking brake being activated.

I stood up and moved to dismount from the LMTV. "Glad you didn't fall asleep, Ranger," I heard from the darkness. It was the Captain. "Move the platoon into the wood line," he explained by pointing to a thicket about 20 meters away. "Administrative patrol base tonight prior to the Santa Rosa Island mission. Don't need you all drowning from lack of sleep. Two lines of Rangers and two up always."

"Roger sir!" I briefed the Platoon Sergeant and Squad Leaders now struggling to unload the exhausted students from the trucks. The news came as a saving grace for the whole platoon who quickly moved into the designated sleeping area. The first two students were assigned watch on the ends and after doing a final count, I fell asleep in the middle of the group opposite the Platoon Sergeant.

"Who is the PL and PSG?" turned out to be our platoon's alarm that morning. We had slept through our 0430 wakeup, and it was now closer to 0530. The new RIs had shown up and were now raising hell. The watch must have fallen asleep! Nobody was awake to get us up in time.

I yelled, "Here I am!" and moved to the RI. "Why is there no guard?" was the first question. "They fell asleep, Sergeant," I explained.

"Who fell asleep on guard?" he asked, this conversation rapidly turning into an interrogation.

"I don't know Sergeant," I told him in a lie. The Platoon Sergeant and I both knew all we had to do was ask who had guard watch by having the platoon raise their hands. Whoever had guard last would then point out whether it was them or the next shift.

"Ranger! Who fell asleep? I want roster numbers!" The whole platoon was watching me now. Every pair of eyes were fixed on me as I was grilled into giving up one of our own. "Sergeant, I take full responsibility for the lack of a guard."

"Shut the fuck up, Ranger! Tell me who fell asleep, or I will put you down as a Serious Observation Report for failing to respond to an order."

"I fell asleep, Sergeant," I coughed out, wanting an end to this conversation. "Bullshit, Ranger, this is your last opportunity."

"I take full responsibility for the actions of the Rangers under my charge, Sergeant."

"Have it your way, Ranger. See your evaluator for a counseling when he returns this morning."

"Roger, Sergeant."

Walking back to the group as a new set of roster numbers was called out for the final mission of the FTX, I thought I had just signed my own death warrant. Turns out though that providence had something else in store.

A couple hours later, the Captain called me over for a counseling on my performance. He explained what went right and what could have been done better on the mission. He sounded satisfied, though unimpressed. He then asked why I hadn't told his NCO who fell asleep on guard. I explained to him, a fellow Officer, that I would rather die than be labeled as a rat and lose the trust of my men.

"Be careful which hills you die on, Ranger," he advised. "However, I would have done the same thing. Good for you for standing your ground." I was released to return to the platoon.

That day was one of firsts for my Ranger School experience and lasts as well. We hit Santa Rosa for a raid by motorized zodiacs and exfiltrated back to Camp Rudder by UH-60 Blackhawk helicopters. The only hiccup for the culminating mission was that the Weapons Squad couldn't get a gun operational. The tie downs from the Type III Nylon (550 cord) on the spare barrel bag were caught on the rounds and pulled into the machine gun, rendering it inoperable. In a comedic display, I witnessed two RI's kicking and fighting a medium machine gun to free the sheared fibers from the weapon but to no avail. Unfortunately, they turned their anger on the poor Weapons Squad Leader who received a "NO-GO" and would end up repeating Swamp Phase.

Touching down back at Camp Rudder early in the morning, there was less than a week until graduation for the chosen few. We filed into our barracks and passed out until 0500 hours when we would have our first hot meal in ten days. After demolishing French toast that would have impressed Gordan Ramsey, we spent the day cleaning weapons and turning in equipment. Peer evaluations, done to determine who lacks the character of a Ranger, were completed at nightfall.

Our last day in Florida was one of plenty and amusement. We were afforded a DOGEX but now with Dominos pizzas and Waffle House pies. I found out I received my "GO" from my second evaluation that afternoon. I was moving on to graduation in five days' time. Mail and care packages were issued to all students that evening. To my surprise, I had a package from my Uncle Jeff. Opening it up in front of the RI who was inspecting for contraband, a single MRE fell out. "Wow, Ranger! Your family hates you!" a black shirted NCO exclaimed. "This poor bastard got an MRE in his care package!"

I called my family and Michelle to tell them I was graduating on November 10th, just one day prior to Veterans Day. I was incredibly happy to know Michelle was going to make it to see me. I found Hess later that evening. "NO-GO man," he told me. I was shocked. He was one of the best Cadets at EKU and a hell of a smart guy. He even graduated Pathfinder School while as a Cadet.

"Keep your chin up, brother," I told him. "You got this man." J.T. Hess ended up graduating right before the Christmas exodus so in the end he earned his tab prior to his first assignment at Fort Bliss, Texas.

The next day, I boarded a bus to Camp Rogers at Fort Benning, only three days and a wakeup until graduation at Victory Pond. I recall as we departed Camp Rudder witnessing the macabre spectacle of the recycled students milling around the camp or doing landscaping. I tried not to make eye contact. One of those students was from our platoon. He was labeled as a chow thief for taking someone else's MRE and the entire squad promptly demolished him on his peers. He had even received his GO. Served him right, I guess.

The bus stopped at a McDonalds, but we were not asked to give our orders. Three RIs bought breakfast for them and the driver. The smell of sausage McGriddles made my mouth water. I looked down at my own MRE for the trip. Maple Sausage Patty Pork, a breakfast MRE. After smelling the Mickey D's, I lost my appetite for this MRE, the meal that had been my favorite during this course but one that I would never wish to consume again.

Our final days in Ranger School consisted of turning in equipment, a class photo, and graduation rehearsal. We were fed breakfast and dinner at Camp Rogers DFAC and were no longer rushed or smoked for any petty offense. Most importantly, we received two passes, one for six hours and one for eight hours on the days prior to graduation.

I used my time taking a real shower, eating like a glutton, and spending quality time with Michelle and my parents who came down to graduation. I remember my mom freaking out when we went to Fuddruckers in Columbus. I ate everything I saw on the table. I couldn't help it due to the 61 days of training I just endured.

"Randy, make him stop!" she kept pleading with my father as I filled myself past the capacity of my stomach.

Graduation Day was one all Rangers remember fondly. After a kick ass demonstration called "Rangers in Action" by the RIs in demolitions, hand to hand fighting, rappelling, and weapons, we were pinned by our loved ones after reciting the Ranger Creed for the final time as a student. My father personally pinned the Ranger Tab on my left shoulder. I completed the toughest training in my Army career at the age of 22 and had only been an officer for six months. 344 Students began with me on September 10, 2021, but only 112 students stood with me on the shores of Victory Pond that day.

But it paled in comparison to what was to come in less than a year. I didn't know it yet but in only three months, I would be in Afghanistan as a part of Operation Enduring Freedom.

I have only one piece of advice for prospective Ranger School students. Focus on the then and now. Concentrate on the task at hand and not what comes later. Too many students quit because they are exhausted or hurt when they know there is plenty more to do. They create their own self-doubt which in turn becomes self-pity. By considering yourself with the present, anxiety about the future cannot take root.

Retired Major Randall Pitcher pinning Josh's Ranger Tab on during Ranger School graduation (NOV 2011)

CHAPTER 6

BE CAREFUL WHAT YOU WISH FOR
DECEMBER 2011 TO FEBRUARY 2012
TRANSITION TO FORT BRAGG

With IBOLC and Ranger School training complete, my time at Fort Benning was drawing to a conclusion. I had Pathfinder School scheduled in January 2012 so I decided to see what additional training I could fit in prior to Christmas leave. As it just so happened, the next Air Assault School class at the U.S. Army National Guard Warrior Training Center began a mere four days after I graduated Ranger School. I decided to enroll through the IBOLC school's NCO to at least know some of the sling load procedures and aviation capabilities in this course. I already had the packing list and could go home every night.

I completely overestimated myself in every capacity. The Zero Day of Air Assault School consists of a modified PT test and an obstacle course through which students were smoked between obstacles. Paling in comparison to the Darby Queen that I completed without incident, the Air Assault obstacle course about did me in for this one factor: I was completely out of shape. Ranger School had sapped my upper body strength and stamina. Although I could ruck like nobody's business, I had to repeat the "Tough One" obstacle, an event where a student climbs a rope about fifteen feet, walks across a log ladder, climbs up another ladder until you are about fifty feet above ground, and then descend on a cargo net. By sheer willpower, I hauled my way up that rope despite not having any feeling left in my arms and shoulders.

The next nine days consisted of classroom blocks of instruction and some PT. Once again, compared to what I just went through, it was a breeze. The sling load hands on exam was stress inducing because of the time hacks but it wasn't anything to lose sleep over. The instructors gave students all the time they needed prior to exams to study. The final phase was all rappelling and a 12-mile road march which I passed without breaking a sweat. The entire course was hyped up to the prospect of rappelling out of a helicopter on the day prior to graduation. However, descending from a UH-60 Blackhawk from 90 feet was anything but culminating after executing three airborne operations that fall. That, and it lasted for a grand total of 90 seconds.

In early December, I stood in my third graduation class in three months. I could now add Air Assault School to my resume. My instructor from Airborne School, SSG Cameron Gemoets personally pinned my Air Assault Wings to my chest before immediately punching the wings into my chest. Known as blood wings, a time-honored tradition now banned in the military, I looked down and later saw two tiny smears of blood on my undershirt where the pins on the back of the unsecured badge had driven into my skin.

During the course, my friend from high school, Marty Miller, a Sergeant in the 4th Squadron, 73rd Calvary Regiment in the 82nd Airborne, called me to wish me congrats on my tab. He also informed me that the 4th Brigade Combat Team was about to deploy to Afghanistan that next spring. My orders had me going to the 3rd Brigade Combat Team which was not deploying.

I had a decision to make, and the consequences have stayed with me since. I called my Branch Manager at the U.S. Army Human Resources Command and asked to be reassigned to the 4th Brigade Combat Team due to their upcoming deployment. Without any thought, he did just that and my orders were ready the next day. I was going to Afghanistan in a couple months. I just now had to tell Michelle who was obviously not pleased. The wedding would now have to wait until 2013 which meant a five-year engagement! Despite the disappointment, we continued to look at the positives. By 2012, Michelle would graduate from EKU so that she could join me at Fort Bragg upon my return that fall. I had no idea that this plan was about to take shape much differently than I had expected only a few months later.

The week after Air Assault graduation, I was on my way home to Kentucky for the first time since May. It was for Christmas exodus. I reflected much on that eight-hour journey along I-75. Since May, I had earned my bachelor's degree in history and was commissioned as a Second Lieutenant in the United States Army at the top of my class.

I qualified as an Infantry Officer at IBOLC and graduated as the Distinguished Honor Graduate. I completed Ranger School without recycling and earned my Air Assault Wings just two weeks afterward. I made some incredible friends along the way, and had my fiancée Michelle as my number one fan. I had Pathfinder School in two weeks and would thereafter move to Fort Bragg, North Carolina where I would be a Paratrooper in one of the most storied units in the U.S. Army.

2011 was arguably one of the most important years in my career because it was not only my introduction to the U.S. Army, but it validated me with the minimum requirements to serve our Nation and lead its sons and daughters in combat.

I spent my time between my parent's home in Rineyville and Michelle's family in Winchester. For three days, one of my best friends, Scott Stafford, and I rented a cabin in Gatlinburg, Tennessee where we took the ladies. Between sampling moonshine, the Dixie Stampede, and soaking in the air of the Smokies, life was good.

I had a swagger of sorts from everything I had thus far accomplished. Scotty was commissioning in next May as a 2LT in the Field Artillery and would also obtain orders to Fort Bragg with the 2nd Brigade Combat Team.

While in the Smokey Mountains, Michelle was insistent to visit every store that procured anything relating to Christmas. She was determined to have everything she needed for her Christmas scene in our future home and had accumulated an enormous haul over the years since we had met. Our future in her eyes was set. The truck I had just purchased upon Ranger School graduation, the house with the white picket fence, her working husband, and the basset hound she would adopt.

Everything she ever wanted for was simple and represented the purity of her. To this day, I still cannot believe how I failed to see just how special she was and how perfect of a military spouse she would make.

2012

The holiday break came and went as they aways do. Although those two weeks were the longest break I had in over a year, it still didn't feel like enough time; especially considering I had barely seen Michelle since leaving EKU the previous year.

Returning to Fort Benning, I was anxious to begin 2012 and wondered what Afghanistan would bring. I had heard nothing good about it and rumors abounded since college that the campaign had winded down the previous few years since the 2003 Invasion of Iraq. My only warning came from my ROTC training at LDAC at Fort Lewis, Washington during the summer of 2010.

One of the best NCO's I have ever met, MSG Brian Disque, had warned our platoon of Cadets of the virtues of taking our training seriously, as there were no second chances for many. His words, "Afghanistan will make you wish you would never have been born," stuck to me like glue. They were some of the truest words I have ever and will ever hear. And the go-getter I thought I was at the time had specifically requested the 4BCT of the 82nd since they were deploying. Be careful what you wish for.

The Ranger School Honor Graduate, another IBOLC 2LT, and I met up with the Schools NCO on the Friday prior to starting Pathfinder. Unfortunately, the report date we were given placed us during the second day of training. As new 2LTs, we had no idea what the Army Training Requirements and Resources System (ATRRS) was at that time and figured we could trust the source. That came back on us hard when we were informed, we couldn't enroll in the course and would have to wait until the next class a few months later.

To add salt to the wound, I received a welcome email from the Deputy Commander of the Fury Brigade. All incoming IBOLC graduates should report immediately to Fort Bragg and in-process as soon as possible. At the time, there was a force cap on the Service Members overseas and there was a risk that we may not deploy at all if we delayed. I could either stay at Fort Benning and delay moving to Bragg for a school, or I could arrive at my future battalion and serve as a Platoon Leader in combat.

Pathfinder School would still be there after I returned from Afghanistan, I figured. After quickly out-processing Fort Benning, I cleared my room at FOB IP. Hitching a UHAUL trailer to my newest purchase, a black 2011 Dodge Ram that I bought while back home for Christmas, I put Fort Benning in the rear-view mirror.

Late on the evening in the middle of January, I pulled off the exit from I-95 towards Fayetteville, North Carolina, home to Fort Bragg, and its tenants, the Airborne and Special Operations Forces.

Fort Bragg is one of the most famous military installations in the United States. You won't find many tanks as there are at Fort Knox or Benning. You also won't find many Soldiers out of shape. Most Soldiers were Paratroopers belonging to the 82nd Airborne Division or XVIII Airborne Corps Headquarters. A large population consisted of the Special Operations Forces tenants such as 3rd Special Forces Group,

Psychological Affairs, Civil Affairs, and the John F. Kennedy Special Warfare Center and School.

The best Soldiers in the World trained here at one time or another in their careers. You would be hard pressed to drive down Ardennes Road or Longstreet without finding someone running and the number of fitness centers on the installation still baffles me to this day.

After two days of in processing, I landed at the footprint of the 4th Brigade Combat Team "Fury From the Sky." The brigade itself was composed of its Headquarters; the 1st and 2nd Battalions of the 508th Parachute Infantry Regiment; 4th Squadron, 73rd Cavalry Regiment where Marty was posted; the 2nd Battalion, 321st Airborne Field Artillery Regiment; the 782nd Brigade Support Battalion; and the Special Troops Battalion. I was assigned immediately to the 2nd Battalion of the 508th Parachute Infantry Regiment (2-508th PIR) "Two Fury!" or 2-Fury.

Out of all the footprints on Bragg, this one was by far the most convenient as every unit was co-located near each other with the DFAC and barracks on site. Tall pine trees shrouded the brigade's units from the rest of the division, creating a secluded community in the middle of the Fort Bragg cantonment area.

I was immediately met by the 2-508th PIR Battalion Operations Officer (S3), Major Mayo who kindly gave me a brief orientation of the unit prior to my first introduction to the Battalion Commander, LTC Guy Jones. MAJ Mayo explained to me about the dynamics of the unit after learning about my first months in the Army at Fort Benning. I could immediately tell he was incredibly intelligent and cared a lot about the members of the S3 team who worked under him.

Soon afterwards, I was interviewed by LTC Jones. LTC Jones asked about my transition and my training at Fort Benning. He shared his vision and leadership philosophy with me and emphasized the importance of character attributes and competencies in his team's leadership. I learned that I would be made into a Platoon Leader in the Battalion's

Weapons Company (Delta Company) that had been reconfigured into a Rifle Company for the upcoming deployment. The Platoon Leader I was to replace would deploy with his Paratroopers to Afghanistan and a couple months later, I would replace him in theater. In the meantime, I would shadow another Platoon Leader in Delta Company until the unit deployed a month later, in mid-February 2012.

I next met my new Company Commander, CPT Brian Bilfulco, who introduced me to the unit's leadership during a meeting that evening. 1LT Pete Kavanaugh was the officer I would shadow for the next three weeks prior to departing. Even though I had trained my ass off over the previous year, there was still so much to learn in so little time. The Company 1SG, Sergeant First Class (SFC) Kelly quickly instructed me to complete in-processing at the Battalion Headquarters prior to moving down to the platoon. He explained that I needed to ensure all my affairs were in order, complete a jump with the unit to remain on pay status prior to deploying, and that I needed to obtain the equipment necessary to deploy to Afghanistan at the Fort Bragg Central Issue Facility (CIF). In essence, I had very little time to waste.

My residence at this time was on a third-floor apartment in a small town adjacent to Fort Bragg called Spring Lake. I lived in spartan conditions during this period and the only furniture I had was a couch and an inflatable mattress. Since internet could not be installed that week, I was forced to utilize a public library to complete the mandatory online courses required of all Soldiers deploying overseas in a combat zone.

Pete was incredibly helpful in helping me assimilate to life in an airborne infantry company. He and his PSG SFC Adrian Ramirez allowed me to attend the training with their platoon in the field and conduct PT with them. Even though I was a stranger, the Paratroopers were very helpful and full of personality. On my first day with them, I was promptly mobbed and "initiated" into the platoon. A few sores and bruises later, I was officially welcomed like a member of their team, even though this wasn't even the platoon I was destined to lead.

The training that Delta Company conducted was realistic and simulated the conditions that the Paratroopers were expected to encounter. We developed and refined TTPs on avoiding IEDs and trained constantly on IED and mine detection equipment that we would operate while patrolling in theater. Combat lifesaving, reacting to contact, tactical callouts, and presence patrols were rehearsed until every Paratrooper understood their roles.

Everyone took the training seriously, especially the officers and NCOs who had been with the unit during the 2009 deployment to Kandahar's infamous Arghandab River Valley where some serious fire fights occurred. "There will be casualties," 1SG Kelly cautioned the Platoon Leaders and Platoon Sergeants. There was no ambiguity about what we were about to walk into. Zharay District was the spiritual birthplace of the Taliban movement and the town of Sangsar within it had been the home of its leader Mullah Omar. In essence, this was the lion's den that we would be patrolling for the next eight months.

A few days prior to the deployment, my parents and Michelle traveled to Fort Bragg to spend the Super Bowl with me and see me off. I had also finally decided to make the decision and proclaim Jesus Christ as my Lord and Savior. This was of immense importance to not only myself, but to Michelle, who arguably put me on the path towards salvation through her grace and character.

During college, I had started attending church with her family every Sunday and was moved by her faith. Just prior to going to war, this seemed like the best time as ever to undertake one of the most important decisions in my life. The Battalion Chaplain, CPT Rob Belton, baptized me that Sunday morning at the Fort Bragg Field Artillery Chapel. My Certificate of Baptism is still stored proudly in my career binder full of all my documents pertaining to my Army life after commissioning.

It was a bittersweet feeling spending those moments with Michelle and pained me to know that I wouldn't see her again until September when the brigade was expected to return. She was nonetheless proud of me and everything I had accomplished. I was of her too, my fiancée finishing her last semester at EKU prior to earning her degree in social work.

Soon after I returned, we would be married. We had a plan and cherished the day when it would come to fruition. If only we knew that we would see each other again much sooner than anticipated. She would not be there at Green Ramp nor the Battalion Headquarters on the day of my departure. Back in Richmond, Kentucky for her studies, this day was on me, although I envied the spouses saying goodbye to their Paratroopers in person.

The morning of my departure, I left my small, unfurnished apartment as if I would be back that afternoon. About the only thing I did was ensure the utilities were shut off and the fridge was empty. Lacking a safe, I sold most of my firearms the previous week and kept only a small Sig Sauer P250 that Michelle had gifted me on my birthday a couple years previously under my pillow. I drove my Dodge Ram to the deployers parking lot and parked, praying nobody would break into it, although there was nothing of value in it.

Delta Company was on another Main Body flight. I was to fly with the Headquarters and Headquarters Company (HHC) of 2-Fury as I was still on their roster and not yet permanently assigned to Delta Company. I linked up with some of the members of the Operations Team under Major Mayo, with whom I would serve under until I took a platoon in a couple of months.

After drawing our weapons and getting accountability, the leadership of our chalk had us board white school buses for the short trip to the PAX Shed at Green Ramp, the Fort Bragg Aerial Port of Departure for Paratroopers and Soldiers deploying. To my dismay, I was issued an M16 Rifle as opposed to the standard M4 Carbine. I was the only member of our flight with one and jokes from the senior NCOs of "My father used that in the Gulf War," or "Nice musket Sir!" would be endless during this journey.

We arrived at Green Ramp and waited for what seemed like hours for the commercial aircraft to arrive. Members of the Patriot Guard, a motorcycle community that famously protects and secures the funerals of fallen Service Members from protestors, were at the PAX Shed in force, handing out coffee, food, and phone cards. I did not know it yet, but these phone cards would be my lifeline to Michelle until I could find a Morale Welfare and Recreation (MWR) facility to call her.

When the plane finally arrived and fueled, we were marshalled again. A quartering party of the Main Body Officer in Charge (OIC) and NCO in Charge (NCOIC) inspected the seating configuration of the plane prior to any other Paratroopers loading. When they returned, the announcement of "Officers and senior NCOs at the front of the aircraft," was announced. I did not think too much of this as a brand-new butter bar. There were a lot of officers in this group and most of them outranked me and all of them had more time in service. Finally, we commenced the walk to the aircraft.

The time was now, the culmination of months of training. On February 15th, 2012, I boarded a commercially contracted aircraft operated by Omni Airlines to the former Soviet Union. A member of the second large main body of the brigade deploying, I surprisingly found myself seated in the beginning of the aircraft where the first-class seating arrangement normally lay. Reclining back, I popped a couple Tylenol PM and put on a sleep mask, praying for a safe return to the States in October. I dozed off among my new comrades, unaware of the dangers that lay ahead in 2-Fury's Area of Operations (AO).

My crew from Truck One of the 2Fury PSD. (MAR 2012)

CHAPTER 7
GREEN ON BLUE
FEBRUARY TO APRIL 2012
ZHARAY DISTRICT, KANDAHAR PROVINCE

On September 11, 2001, Al Qaeda hijacked four civilian airliners in what was to become the worst terrorist attack on American soil and the biggest loss of American lives on U.S. territory since the Japanese attack on Pearl Harbor in 1941. Two aircraft slammed into the World Trade Center in New York City, one into the Pentagon in Washington D.C. and another that tragically crashed in Pennsylvania after the passengers heroically fought back.

The United States' retribution was swift with the deployment of Special Operations Forces and subsequently conventional forces that overthrew the Taliban, thus denying Usama Bin Laden and his cronies' sanctuary and training grounds in Afghanistan to launch further attacks. Now, more than a decade later, it was my turn to do my part. I was in seventh grade at James T. Alton Middle School in Vine Grove, Kentucky when America was attacked on September 11, 2001, and I always knew thereafter that my destiny lies in defending my homeland.

After a brief layover in Manas, Kyrgyzstan for three days, I boarded an Air Force C-17 Globemaster for the two-hour flight south to Kandahar Airfield (KAF). Wearing my body armor, it seemed to finally hit home that this was happening. I was going to war. After being raised in military communities, commissioning as a 2LT after four years of ROTC, and a year of intense Infantry training at Fort Benning, I was now on my way to the heartland of the Taliban.

As I disembarked from the plane that morning in Kandahar Province, a grim landscape greeted me. A couple of high-rise mountains loomed over the city of Kandahar, but otherwise the urban center looked as though it had seen better days. This was just a glimpse, as I was in the middle of one of the largest U.S. bases in the country, surrounded by tarmacs, hangers, and facilities maintained by Ecolog and Kellogg, Brand, and Root (KBR). A steady rain began falling as the flight OIC shuffled the manifest of over two hundred Paratroopers to a large tent to be processed into the country.

When we reached the transient barracks in the late afternoon, over three inches of standing water greeted us. Over a hundred other Soldiers were occupying this giant circus tent lined with row after row of double stacked military bunkbeds. Gear and bags hung precariously off the bunks to remain dry and the stench of body odor and stale air lingered. This was our temporary billet for the next three days before departing Kandahar for the Forward Operating Base (FOB) where the Battalion Headquarters was stationed.

I explored the confines of the sprawling base with a couple other officers from the Battalion Staff. I was amazed at the amenities available to U.S. and NATO Soldiers on the installation. Although the typical destinations such as a Post Exchange, an MWR, and a coffee shop called Green Beans was on site, what I witnessed next astounded me. KAF had a boardwalk that was built around a small soccer field. Surrounding the wooden accessway was a series of restaurants, stores, and parlors. T.G.I Fridays, KFC, Pizza Hut, electronics stores, carpet merchants, massage parlors, and barbers lined the boardwalk.

It was mesmerizing to think Soldiers could have these luxuries while deployed. However, for the Paratroopers of the Fury Brigade, these amenities for the most part remained out of reach, the bulk of the brigade pushed out to stations around the province.

A couple days later, I was standing on the tarmac waiting for an aircraft to pick us up. Due to the prevalence of improvised explosive devices (IEDs) on roads outside the perimeter of NATO bases, air travel was the quickest and safest means of getting from place to place. Two enormous CH-47 Chinook helicopters swooped in to pick up the group I was traveling with. The dust from the ground swept away in a large drift as the rotors from the bus sized aircraft whirled overhead. Upon touchdown, we were motioned by a crew chief to approach the rear of the aircraft. With my ruck and duffel bags, I carefully took a step onto the tail ramp and settled as far forward to the cockpit as possible to create space for the remaining Paratroopers.

Although the flight couldn't have lasted more than thirty minutes, it felt much longer. The late winter weather was surprisingly sunny and mild, but I do not recall taking in much of the landscape due to my seating position on the Chinook. Our destination was FOB Howz-E-Madad, the headquarters for Task Force 2-Fury. Located just north of Highway One in the middle of Zharay District in Kandahar, the base was only a few kilometers from where Mullah Omar had launched his offensive against the warlords' decades earlier.

Two terrain features encompassed most of the AO, the Arghandab River Valley to the south, where the battalion had fought during their 2009 campaign, and to the north flanked by Highway One.

Touching down on the hard stand within the base, we were driven first to a small tent that served as the Mayors Cell. We were quickly issued our combat supply of ammunition for our M4 Carbines and some additional first aid dressings. By this time, I had exchanged my M16 for a standard M4 with a PEQ-15 laser and M68 optic.

Soldiers from the 1st Battalion, 32nd Infantry Regiment (Chosin) of the 10th Mountain Division from Fort Drum, New York helped receive us onto the base and oriented the newcomers to the layout of Howz-E-Madad. These Soldiers made up the Task Force that we were replacing. Already a sizeable portion of their battalion had redeployed and the remainder served to train and help us assimilate into the battle rhythm prior to their departure.

After dropping my duffle bags off into a tent that most of the other staff Lieutenants occupied, I secured my weapon and made my way to the Battalion Headquarters located in the center of the FOB. This location served as the location of the Battalion Tactical Operations Center (TOC), the offices of the staff, the conference room, and the command team suite.

Faces I had briefly met at Fort Bragg appeared as I explored this new structure. Perhaps the biggest surprise was my first introduction to the various Soldiers of the Afghan National Army who worked on the base within the TOC as translators and liaisons. So far these were the first Afghans I had met and most seemed too busy to deal with a young green 2LT who had just arrived at their homeland.

Still awaiting to take a platoon in D Company, I was assigned to be the Battalion Command Team's Personal Security Detachment (PSD) Platoon Leader (PL). With four heavily armored Mine Resistant Ambush Protected (MRAP) vehicles mounting a lethal combination of M2 .50 Caliber Heavy Machine Gun, MK-19 Grenade Launcher, and M240 Bravo Medium Machine Guns, I had a squad sized element of Paratroopers from HHC under my leadership.

I rode in the first vehicle with my driver, SFC Bradley Kelso, our gunner SPC Brian Errickson, and Female Engagement Team Leader, SPC Kelsie Tate. The second MRAP carried the Battalion Commander, LTC Jones, was driven SSG Patrick Kelhi, and gunned by SSG James Ridenour. The Commander's personal guard was SGT Joshua Parker. The third vehicle escorted the Battalion CSM, CSM Steve Green and was driven by SGT Daniel Ruggiero. The trail MRAP carried our Platoon Sergeant (PSG) SFC Sean Lee.

Our focus during operations was to transport LTC Jones and CSM Green throughout the AO to different Combat Outposts (COP), each operated by a company within the task force. Otherwise, we would make visits to FOB Pasab about a dozen kilometers east of FOB Howz-E-Madad where we met with the Brigade Commander, COL Brian Mennes. Most of the visits concerned visiting Paratroopers and leaders, or Key Leader Engagements (KLE) with the local tribal and village elders within Zharay District. When we arrived at our destination, our dismounts escorted the command team while the vehicles pulled security outside of a compound.

When not conducting operations, our team was trained daily by SFC Sean Lee. Marksmanship, vehicle maintenance, and recovery drills typically were executed on days when we had no missions while I planned routes and made coordination's for linkups in the AO. For a new officer, the PSD gig may not have been as sexy as being a Rifle Platoon Leader, but it had some advantages.

For starters, I was able to see the entire AO and the Brigade Headquarters (BDE HQ) during my missions, thus gaining insight on the local terrain, the capabilities of our Afghan partner force, and the dynamics of the Pashtun population within Zharay District. In addition, the daily Battle Update Briefs (BUB) and Commanders Update Briefs (CUB) cued me in on the Battalion mission, the Command Team's priorities, and the significant activity within the AO.

After two weeks of this duty, the reality of combat hit me for the first time. I was awakened early by SFC Lee that an attack was reported on the Bravo Company COP located in the small village of Sangsar a few kilometers south of FOB Howz-E-Madad. The attack had been perpetrated by members of our Afghan partner force whom our comrades had known well. Known as Green on Blue engagements, these attacks occur when Taliban insiders of the Afghan army and police forces attack a U.S. position or Soldier.

At around 0300 hours on March 1, 2012, one of the Afghan NCOs approached the COP Sangsar Entry Control Point (ECP) and shot a Paratrooper in the chest, stunning him before alerting his accomplice, an Afghan schoolteacher on the base. They then crept up to the nearest guard tower and shot SPC Payton Jones in the back of the head as he scanned for any targets after hearing the shot. The Taliban insiders then turned the medium and light machine guns on the base and fired Rocket Propelled Grenades (RPGs) into the COP.

The Sergeant of the Guard, SSG Jordan Bear, was fatally shot as he responded to the gunfire at PFC Jones's position. At this point, the Paratroopers of Bravo Company understood the situation and zeroed in on the occupied tower with a hail of 5.56 and 7.62 under the leadership of 1SG Hissong. The attackers shot an RPG at the mortar pit, disabled a MAT-V trying to maneuver on the tower, and set multiple fires by rupturing fuel lines. It was a full-scale firefight in the confines of this tiny COP.

Air Weapons Teams (AWT) consisting of two AH-64 Apache Gunships were called in for support in case the attack was much larger in scale. Since the tower was too close to the base, the pilots couldn't risk firing Hellfire Missiles without potentially injuring friendlies. The traitor's downfall, however, was the smoke from the spreading fires which began to collect in their fortified position.

They egressed out of the tower window and dropped roughly two stories to make a break for it. However, the Apache pilots, now with clear shots and armed with their night vision capabilities, killed the intruders before they could get very far. The cost of the attack resulted in two killed in action (KIA) for Bravo Company during their first month in theater. The Paratrooper initially shot in the chest was protected by his armor and was instrumental in helping identify the insider who shot him. The mortar firing pit was destroyed, and numerous vehicles sustained damage. Trust between the Afghan forces and the Paratroopers was severely strained, the faith that our partners had our back now shaken to our cores.

The attack had just ended when the PSD pulled up to the base. An Apache buzzed overhead, less than fifty feet off the ground, as a show of force to thwart any follow-up attacks. As LTC Jones and CSM Green were briefed by the company leadership on the events that had just unfolded, I walked around the camp. The guard tower was severely damaged from the gunfire and debris strewn everywhere. Bravo Company did not fall into despair though, but instead went to work as they had been trained. The unit set about immediately repairing their positions and increasing security. Although they remained stoic after this fight, I could tell it left an impact on this team.

Just two weeks later, I was informed by LTC Jones and CSM Green that they were confident and comfortable with me going down to D Company and conducting a handover with the outgoing Platoon Leader, 1LT Brian Yoder. Packing my bags, I said farewell to my first team I ever had in the Army. My replacement, 2LT Adam Brown, was taking the reins from me while he waited for his turn to take a platoon. He did not have to wait long.

I was picked up by MRAPs from Delta Company's First Platoon and driven to a platoon strong point off one of the main avenues of approach throughout DCO's (abbreviation for D Company, used going forward) battlespace. The location was a square compound built of HESCO barriers and manned on the corners by the platoon's MRAPs. A large tent served as a living space for the Paratroopers and a smaller tent served as platoon command post. Spartan and austere may be the best words to describe the conditions of this compound, known as Strongpoint (SP) Sartek.

SP Sartek had no showers, no dedicated DFAC, and an open-air toilet with some tubes driven into the soil served as the latrine. Every day, a Paratrooper was assigned to burn the foul contents with JP-8 fuel. Squads daily patrolled the surrounding villages while another squad drove around in the periods of darkness to look for signs of Taliban activity, known as Lines of Communication or LOC patrols.

For two weeks, I integrated with First Platoon at SP Sartek through-out the DCO AO within Zharay District. The Company HQ and Sec-ond and Third Platoons operated out of the unit base known as COP Zharifkel, a few kilometers from Sartek. Two aspects concerned my dai-ly existence. First, my priority was to conduct a handover with 1LT Yod-er before he transitioned out of the Platoon Leader position and moved to FOB Pasab to work with the BDE HQ. This entailed attending all briefs with the outgoing leader, conducting inventories of the platoon weapons, vehicles, and equipment that I would sign for, and learning the capabilities of this platoon of Paratroopers.

Next was the operational aspects of life in Afghanistan. Squads ro-tated on dismounted patrols during the day or moved by tactical vehicles by night. SFC David Deal, the PSG, took ownership of the LOC pa-trols at night while 1LT Yoder and the Squad Leaders patrolled by day. The Squad Leaders were SSG Abt, SSG Ansari, SSG Lewis, and SSG Holland, experienced and competent NCOs who were well-respected within the platoon.

Everyone dreaded the LOC patrols throughout the AO, missions that deterred insurgents from emplacing IEDs or attacking U.S. or Af-ghan forces through a continuous presence. They left the strongpoint in the evening and returned early in the morning after a long night of driving up and down Zharay District while stopping at FOB Howz-E-Madad for fueling. They were dangerous and the crews had to remain awake when they returned to complete various tasks around the outpost.

The dismounted patrols consisted of a squad and a gun team moving from Sartek to pre-determined locations near the strongpoint. Typical-ly, the patrols either conducted KLEs with locals or simply executed presence patrols throughout villages and investigated for signs of IEDs, weapons caches, or other indicators of insurgency activity.

Since the land was infested with IEDs and mines, the patrols had to move in single file behind a Paratrooper operating a piece of equipment known as a Mine Hound. Like a mine detector on steroids, the Mine

Hound could detect metal objects in the ground but also had ground penetrating radar that could pick up other objects or disturbances in the Earth. It was a solid piece of equipment and was a pacing item for the unit which no doubt saved many lives.

In essence, the squads of First Platoon had the following rotations: LOC patrols, dismounted atmospherics patrols, and base security. If Paratroopers weren't doing one of these missions, they would be executing one of the other two. There were no days off. On hours between shifts, they spent their scarce time either sleeping, eating, doing hygiene, or lifting some of the weights that had been brought over by the unit.

Since there were no showers, the only way to stay clean was to use baby wipes. Water was a precious commodity that couldn't be wasted for anything. Paratroopers on LOC patrol or conducting visits to COP Zharifkel could steal away for a half hour to shower but only when time wasn't pressed.

A couple times a week, 1LT Yoder and I would move by MRAP to COP Zharifkel for a Commander's Touchpoint with CPT Bilfulco and 1SG Kelly. These meetings synchronized the platoons in DCO and ensured that all operational and administrative tasks were completed. I was able to see 1LT Pete Kavanaugh of Second Platoon and the Third Platoon Leader, 2LT Eugene Connors, a giant Irishman who hailed from Brooklyn and had the accent to accompany it. Zharifkel had the amenities that Sartek didn't. While still considered austere by the rest of the Army, the COP at least had Porto-johns and a DFAC tent to eat in.

Towards the end of March, First Platoon was tasked to switch out with 1LT Kavanaugh's team and relocate to Zharifkel. Lifts of Paratroopers and vehicles moved out of the strongpoint and were quickly replaced by Pete's men on the perimeter. While the company base was more comfortable and had much more space, Sartek was away from the flagpole. To everybody, this outpost was most preferred due to the autonomy that it offered to the men.

Nonetheless, within a month, they would return to Sartek and patrol their AO again. For now, they moved to Zharifkel to conduct the same missions but with a new environment to observe with different routes and villages to patrol. Since I now had access to internet in the command suite, I could now finish any online administrative tasks that 1SG Kelly or the XO, CPT Farrington, needed me to accomplish.

On March 29, 2012, while knocking out some training within the command suite, a large sudden "BOOM!" and accompanying vibrations startled me. I looked immediately at 1SG Kelly. At first, I thought it was discovered IEDs and unexploded ordinance being blown in place by an U.S. Explosive Ordinance Disposal team, a common occurrence throughout the area. However, the lack of a warning and the look on 1SG Kelly's face told me this was not ordinary or planned. Following my 1SG out the wooden door, I looked to the perimeter wall near the motor pool and saw a large plume of black smoke rising. What I witnessed next still haunts me to this day.

From the other side of a HESCO wall dividing the ammunition holding area from the motor pool, I witnessed Paratroopers limping back towards the basic aid station next to the command suite that I had just vacated. 1SG Kelly and I were among the first on the scene and as he ran forward, I sprinted immediately to a wounded Paratrooper. Screams of pain and shouts of "Medic!" echoed through the air as the base now began swarming with activity.

I went straight to the first hurt Paratrooper, SPC Lee. His legs were riddled with lacerations, and he was profusely bleeding from both limbs. Slinging him over my shoulder, I carried him to the aid station and laid him on a litter. In hindsight, this was a mistake that experience had not taught me.

As the medics were helping those with critical injuries, I immediately began applying two CAT tourniquets to SPC Lee. He was in a lot of pain as bright red blood squirted on my uniform and covered my hands. After securing the tourniquets, I applied pressure dressings to his legs immediately over the lacerations. This was the furthest extent of my Combat Life Saver training that I could apply.

Soon, more wounded Paratroopers were carried into the aid station. A medic quickly came over and triaged SPC Lee. "He'll be fine, we need the bed." Obliging the subject matter expert, we moved Lee to the floor against the aid station wall. There were more casualties than the aid station could handle, and the medics could support by themselves. Known as a Mass Casualty or MASCAL situation, the litters needed to be utilized for the seriously and critically wounded. At this point, all of the wounded were being addressed by Paratroopers and medics alike.

Observing the situation, I tried to get a handle on what was happening. I was not prepared for this. Even after months of training, the screams of the wounded, the shouts of the medics, and the movement of the men around me was a crescendo of action in a moment of chaos. Despite the tragedy, the unit was functioning exactly what it had trained for. The wounded were brought from the point of injury to the aid station. The senior medic and 1SG had triaged them and assigned patients by severity of their wounds to litters or areas of the aid station. Looking back at it, DCO had trained well for this, and the repetitions paid off.

The cause of the event was unexpected to the unit. Improperly stored ordinance left by the outgoing unit detonated as a squad from Third Platoon was getting ammunition to conduct a patrol. It was nothing on the platoon or unit and nobody in the Company could have prevented this.

Despite the cause, consequences ensued. Over a half dozen Paratroopers were wounded on this day, to include the Third Platoon Sergeant. One man, a fellow Kentuckian from Henderson, died as a result. I had briefly met SPC David W. Taylor just prior to this day, calling him out after hearing he was from the Bluegrass State. His demise was the company's first during the deployment. To this day, his name is the only one I wear on my wrist, a reminder of the Heartland's sacrifice.

Two UH-60 Blackhawks came to pick up the wounded and SPC Taylor that afternoon. That was a hard evening for me and the men of the Company. As CPT Bilfulco and 1SG Kelly talked to the platoon leadership that evening, we reflected on that day's events.

Despite the tragedy, the Company continued operations as if nothing happened. After the memorial ceremony, we needed to continue doing our job. Unlike civilian life, our duties required us to patrol and place ourselves in places of high danger. We simply had to carry on. No amount of training prepares you for it. But when the brother or sister next to you relies on your presence to keep them safe, you answer the call and move out. That is the beauty of combat. No matter how much you go through or witness, you report the next day because your team counts on you to do it.

Task Force 2-Fury was full steam ahead on operations throughout April upon the conclusion of the Transfer of Authority (TOA) Ceremony with the departing 10th Mountain Division team. Paratroopers from every outpost throughout Zharay patrolled day and night to identify and reduce the Taliban presence throughout the AO. I executed some of my first named operations with DCO and my time to take First Platoon was at hand. In less than a week, I would officially be leading my platoon and responsible for the Paratroopers' welfare. The inventories were about complete, and I had met every member of the platoon.

On April 14, 2012, Task Force 2-Fury planned an operation to clear a village that up until then had not been patrolled. DCO was tasked as the clearing element for this mission. Kotizi looked on a map or Google Earth like nothing out of the ordinary. But contrary to first impressions, the village had been a haven for Taliban fighters moving from Pakistan to their assigned locations throughout Afghanistan. The unit before us had stopped patrolling there and simply put up a large barrier of concrete T-walls between the rest of Zharay and Kotizi. It would not be a safe area by any means and the Taliban had plenty of time to prepare for our inevitable knock on their door.

Third Platoon was tasked with clearing a route from one of the main roads directly to the village for the future construction of an outpost.

First Platoon was tasked with moving in dismounted and systematically clearing the village. Adjacent units would help with the cordon of the village to prevent unauthorized personnel from coming in and out. It was the largest operation of Task Force 2-Fury yet. We would be clearing from the western most point of the village towards the east where we would link up with other units.

The operation itself went smoothly. The village was searched by Afghan National Army and police forces along with their U.S. counterparts. An 84mm recoilless rifle, various small arms, and a considerable amount of ordinance were hauled in that day. What shocked me the most was the sheer amount of IEDs located by the Afghan Soldiers and our Air Force K-9 attachment led by Air Force Staff Sergeant Donald Nachand and his beautiful German Shepard Jimmie. There had to be over a hundred IEDs found throughout the village. Enough that our Army EOD team had to resort to dumping the homemade explosives in a small wadi (water filled ditch) when they ran out of equipment to reduce the IEDs by blowing them in place (BIP). One of their BIPs, the explosion was so big that the windows in many structures were blown out. The K-9, Jimmie, had to have been exhausted by the time the evening fell.

Perhaps most striking was that the Afghan Soldiers seemed to know where the IEDs were located. I remember watching in disbelief as the Afghan forces pulled them out of the ground like carrots and made a bee line to the next IED to remove. Either experience had shown them how and where their nemesis had employed their main weapons, or they were simply lucky. No casualties among ourselves or our Afghan partners occurred that day, to my knowledge. It was an awesome operation and the battalion had really pulled off a coordinated clearance of the entire village with absolute precision.

That evening, a squad from First Platoon was tasked to conduct an ambush on the southwestern outskirts of Kotizi in the event insurgents decided to surprise us overnight. I volunteered to go, hoping to get some additional experience and give 1LT Yoder a break.

As we departed Kotizi in single file, we moved under cover of darkness and with our NVGs to our position on the bank of another wadi. Although the movement was less than 300 meters in total length, the effects of that day's mission, the darkness, and the slow movement of the squad when following the mine hound, made it feel as though it took forever.

Finally, we arrived at a fine location and assumed positions in a linear fashion. Somebody was kept on watch and the remainder of the team rested in shifts. Placing in a pinch of Grizzly Wintergreen, I stared up at the stars. The night sky in Afghanistan outside of the city was simply amazing. Soon, my body and mind caught up with me and I was able to get some rest prior to the next day. It had been one of the more memorable days of the deployment so far.

My brother and I reunite at FOB Pasab. Last pic with 2 legs.
(MARCH 2012)

CHAPTER 8
THE CATALYST
APRIL 15, 2012
OUTSKIRTS OF KOTIZI VILLAGE

I awoke prior to the sun coming up. Although I must have gotten less than two hours of sleep, I did not feel fatigued. About ten Paratroopers and a two-man Army EOD team were stirring and preparing to reassume their positions on the line overlooking the wadi and some fields of poppy and grapes. Putting my armor and helmet back on, I observed the surroundings as the sun rose and warmed my face. It was quiet in the vicinity, despite the actions of yesterday's operation. I do not recall hearing anything in the area despite the large influx of U.S. forces a hundred meters away. Upon checking my Casio G-Shock watch, I noticed it had died that morning. I must get a new one on my next visit to Howz-E-Madad.

The morning sky was simply beautiful. The temperatures were in the low seventies and the blue sky contrasted sharply against the light brown and tan of the Zharay District's orchards and grape huts. As we were only in place for the duration of the evening, we prepared to exfil back to Kotizi's village where the rest of the force resided. The call of nature had hit me hard that morning. However, there was nowhere else to go.

Severe abdominal cramps hit me. The result of a dinner and KLE with our partner forces the previous evening before the mission kicked off. The local food never agreed with me, but I felt obliged to partake in order to show respect.

Letting the patrol know, I moved about five meters linear of the group to the edge of the wadi. Nothing seemed out of the ordinary and the vicinity had already been cleared. Taking care of business, I observed a white piece of cloth in a tree, nothing out of the ordinary since garbage was everywhere in Afghanistan due to a lack of sanitation facilities. A wise man later told me that you can tell when a country is slipping into Third World status. You can have two options out of three choices: clean water, education for your children, or garbage pickup. In Afghanistan, you had none of those. I climbed out of the wadi and onto the path to return to the group only meters away. They were the last steps I would ever take with two feet.

About five meters from the point man, I heard a sudden and a very loud "POP!" A sudden flash of heat and dust engulfed me for a second and subsided as quickly as it occurred. I took one more step and hit the deck on my backside. I was wide awake. I never passed out and the daze was incredibly brief. I was fully alert, almost ready to engage the enemy as adrenaline had taken over. What the heck was that? A land mine? An IED? Adrenaline quickly morphed into shock as the realization set in. I had all of my kit on, which probably saved my life. The smoke quickly cleared, and the dust settled.

Immediately, I was surrounded by Paratroopers who immediately went to work. The platoon medic, SPC "Doc" Burgos, used his trauma shears to remove the pants covering my leg to check for bleeding from the extremity. He immediately applied two tourniquets on my upper left thigh to stop the bleeding. Bright red blood was everywhere. Another was holding his hands over my eyes. The Squad Leader was calling for the 9-Line MEDEVAC.

"Let me see!" I demanded, the dirty gloved hands obscuring my ability to witness the moment. Pulling his hands up, I saw what was wrong. My left leg at the calf looked like a pretzel. My left foot was completely twisted and dangling, the Rocky brand combat boot over my foot peeled away like a mutant banana peel. I did not understand it at the moment, but a small IED had shattered my left foot and ankle. As the adrenaline and shock wore away, the pain set in. And let me tell you this hurt like hell.

I tried like hell to keep from screaming, lest a local hear the agony of an infidel American. As Doc went to work, lacking any pain killers, I breathed heavily and fast, trying desperately to control my heart rate escalating sharply due to the searing pain. All I know is that I did not cry.

The moment remains to this day so fresh in my head. A unique dichotomy all on its own, the serene peaceful sunny morning now contrasted to the internal chaos I was fighting within me as I tried to comprehend how I arrived at this moment. I didn't pray to God. I didn't think about Michelle. I didn't think about home. At no point did I feel like my lifespan was nearing its end. This is the essence I guess, of being caught up in the moment.

Out of nowhere, the Battalion's Physician's Assistant, CPT Domino, showed up. He had made a bee line from Kotizi straight to me without regard for his safety. He immediately stuck me with a shot of morphine through my right hand. As the euphoria kicked in, the shock now evolved into nausea. The team removed my helmet, armor, and secured my weapon as they prepared me for evacuation. "You'll be walking again in no time," one of the EOD Soldiers told me. "Yeah, they should be able to save that," the other chimed in. They had seen this time and time again in their line of work. They knew my fate, but their false proclamations gave me strength, which I desperately needed.

The sound of a UH-60 MEDEVAC Blackhawk replaced the cacophony of the medics and the Paratroopers preparing to move me out. The bird landed in the field immediately behind our position. Dust kicked up and a brownout partially obscured our vision for a couple seconds. Carrying me by a pole-less litter to the bird, the guys rubbed my shoulder and told me to "take care," and "you'll be back soon." I still struggled to comprehend exactly what just happened, but I did know one thing, I was probably not returning any time soon.

The flight to Kandahar Airfield was lonely. The crew chief manned the M240-B and kept watch on the environment around the aircraft. Even though the flight was less than twenty minutes, to me, alone on a litter and in an incredible amount of pain, it felt like an eternity. As the bird came in for the landing, the crew chief came over and checked on me in preparation for the handover to the Role III Military Medical Treatment Facility at Kandahar Airfield.

I was put on a stretcher and wheeled straight from the MEDEVAC bird to an operating room. Nurses and aids quickly removed any of my remaining gear and clothing. The surgeon quickly observed my leg and let me know everything was going to be alright. My eyes struggled to adjust to the brightness of the operating room. Lights focused in on my body like a concert and I was currently the star of the show. A mask of oxygen was placed over my mouth and another shot was administered. Within seconds, the euphoria of anesthesia overwhelmed me. I collapsed into unconsciousness and forgot every care in the world. Being a Platoon Leader, my fiancée Michelle, the Paratroopers back in Kotizi. None of it was relevant anymore as a black void filled my mind.

God, whom some have told me was absent that day or when sneering at my belief despite my injury, had indeed been with me and those Paratroopers. The IED had partially detonated. A full detonation would have killed me and wounded even more Paratroopers. "Doc" Burgos had been on it with the treatment. The Battalion Physician's Assistant committed an act of valor to reach me. The MEDEVAC bird arrived and whisked me away without incident. Indeed, God had been with us.

I awoke slowly. Every single fiber in my body felt as though it was under a ton of stress. Just raising my head proved a chore as I came to find out the surgery's outcome. For starters, I had IVs going into my right arm. Morphine, saline, and antibiotics were being pumped into me to negate the burning pain I felt and reduced the chances of an infection. Next, I saw a plastic bag dangling from my bed, the tell-tale sign of a catheter in me. I had an oxygen tube pumping fresh air into my nose and I had adhesive pads stuck all over my chest to monitor my vitals. Looking, down, I could see I was completely naked with a sheet covering me, by burial shroud I reckoned.

Then, the final revelation hit me like a tsunami. Looking towards my feet, I saw one hill rise at the end of the bed, where two should have been to form a saddle. Without considering it, I lifted the sheet in desperation. Where my left foot should have been was now a swollen bloody mass of gauze where my ankle and foot had been only hours prior. A tube was protruding out the bottom, a wound vacuum that sucked out all the debris and liquid to promote healing and reduce infection.

So, it wasn't saved I realized. I looked around the room needing to ask questions to anyone who would listen. I must have still been in Kandahar; I wasn't out long enough to be flown out of the country. Not long after this realization, a nurse in camouflage scrubs entered the room to check my vitals. "I need to talk to my fiancée," I asked her. "I need a phone." "I'll find you one," she said.

I dialed Michelle's number slowly. The grogginess of the narcotics made doing even the simplest tasks difficult. "Hello." I heard her voice for the first time in over a week. She had obviously been crying and was in despair. News travels fast I thought.

"It's me." I stuttered.

"Josh!"

"Hun, I'm going to be alright. I'm on my way home," I said, trying my best to reassure her.

"They told me you might not make it. That you were badly hurt by a bomb," Michelle said in a sorrowful tone. "We were told to get to Germany as quickly as possible if I wanted to say goodbye."

"Where are you?" I asked.

"I'm on my way to the airport."

"Cancel that flight. Its only my left foot that you won't see again," I spoke. "I love you babe."

"I love you too," she said, bawling her eyes out thousands of miles away.

"I'll call you when I'm out of the country," I explained, saying good-bye to her. No doubt this phone call was costing the government hundreds of tax paying dollars.

I then called my parents and explained the situation in detail. I did not want them making any knee jerk reactions either. But just try to stop my mom when she is determined to execute a task. Good luck with that.

I was told by a doctor sometime later that I was being moved to Bagram Airfield (BAF) and subsequently transferred to a flight to Landstuhl, Germany, not far from where I was born, to be stabilized and await the next rotator to the United States. It would be at least three days prior to being back in the United States, but for the meantime, all I could do was suck on morphine and wait.

The Deputy Commander (DCOM) of the 82nd Airborne Division, Brigadier General (BG) Sinclair stopped by that afternoon. He placed a Purple Heart Medal on my chest and offered his support. It was a kind gesture, but the Purple Heart was not a medal worth receiving. It is earned by your own blood. Sacrifice if you will. It should have been the enemy's blood on our boots, not my own.

"Is there anything I can do for you," BG Sinclair asked. "I want to see my brother," I said. "He's in the 5-20 Regulars of the 2nd Infantry Division. His unit is operating east of Task Force 2-Fury in Kandahar." This request must have seemed a stretch to ask, I may have thought. But if someone can make something happen, it's an officer with stars on his chest or shoulder boards.

Looking at what must have been an aide, Sinclair ordered him to "find SPC Pitcher and bring him here." The aide stuttered out something about my flight to Bagram or something along those lines. "Get his brother to BAF," he growled.

"I'll make it happen," the aide said. Must be nice to have that kind of influence and power, I thought.

Prior to the DCOM departing, I thanked him and saluted as best I could notwithstanding the tubes poking out of me. "All the way," I said holding up my right hand to my temple. "Airborne, Josh, heal up," the DCOM replied, returning the salute.

I must have passed out on the flight to BAF. I do not recall whether the flight was a UH-60 or C-130 although it must have been the latter with the number of wounded in Kandahar. All I recall next was waking up in a ward full of wounded Service Members. I was heavily medicated for the flight and the grogginess hit me again as it did in KAF.

But to my shock, there he was, my brother Justin, an 11B who had been in country for almost a year. Although we never got along, seeing him was a breath of fresh air amidst one of the worst days of my life.

"What the hell?" he asked.

I explained the situation and that I asked a Flag Grade Officer to snatch him up from the shithole we operated in. He retorted that his 1SG personally had him report for a flight to KAF and then Bagram to see me. A General's request. My brother and me had taken a picture at FOB Pasab only weeks earlier. It is the last photo of me with two legs.

The visit was short lived, however, as not more than a couple hours later, I was on my way to Landstuhl in a C-17 loaded with wounded Service Members. Thankfully though, my brother was due to take leave soon and he would receive two weeks away from the fight to rest and reset himself. It was well-deserved because the deployment had taken a lot out of him. Sadly, he never truly recovered mentally from that deployment and left the Army shortly after.

Landstuhl looks like an ordinary hospital from its interior, just full of medical care providers in a uniform. I shared a room with two wounded Navy Seals who had gotten it in Eastern Afghanistan's Kunar Province.

They were just as bad off as I was, but their sense of humor was well received. I received additional surgeries to debride and treat my wounds. My left eye suffered partial blindness and I was now almost completely deaf in my left ear. I had shrapnel wounds on my legs and groin, and I was now concerned that I may never have children, a dream of both mine and Michelles.

Morphine became my best friend, particularly the derivative Hydromorphone or its brand name "Dilaudid." I had a button which I could push which administered predetermined amounts per hour as my pain increased. When the pain became out of control, a nurse shot a direct dose into my vein for immediate relief.

Despite the pain management, my despair began to rise. I was only 23 years old. I hadn't even been in the Army for one year. I felt like a total failure. I felt as though I let down Michelle, my family, my Chain of Command, and those Paratroopers. I withdrew into a state of self-pity and depression. Despite the quality of the food, I was not hungry and for days, I didn't eat. And as the weight loss continued, my muscles atrophied. I didn't care anymore.

Three days later, I was on a plane to Andrews Air Force Base in the Capital Region. My destination was Walter Reed National Military Medical Center in Bethesda, Maryland. Landstuhl, although vastly superior in medical treatment facilities than Bagram, could not help me anymore.

I wanted to start rehabilitating immediately. I wanted to return to the fight as quickly as possible. I somehow managed to get on Facebook from one of the SEAL's iPads in my room. I changed my status to "I will be back." That's all I needed to say. However, I never would have guessed how much work it was going to take to don the maroon beret of a Paratrooper again.

The C-17 carrying me had other Paratroopers on board from Task Force Fury. One of them, SSG Travis Mills, laid next to me. An IED had severed both arms and legs. As I lay there in my own self-pity, I heard his screams of pain on that flight as medical staff worked feverish-ly to contain his pain. Hearing those cries immediately sparked some-thing in me. I needed to be strong. It could have been a lot worse. I had a papercut compared to many of my other comrades. Even though losing any part of a body is damn traumatic, somebody, somewhere is doing a hell of a lot worse than you.

The giant hold of the C-17 was lined with litters, patients, and scur-rying medical providers who worked tirelessly to control our pain. The sounds of the wounded surpassed the roar of the engines taking off. Soon after we were airborne, a doctor checked my vitals and admin-istered enough narcotics into my body to put down a Thoroughbred. I passed in and out of consciousness, often being awakened by the screams of the many wounded Service Members around me. Every now and then, a nurse checked my vitals and moved on. About halfway across the Atlantic, more Dilaudid was injected into me, and I fell back into a trance.

After the plane touched down at Andrews Air Force Base outside of Washington D.C., its human cargo was delicately transported to a specialized patient bus. Removing my oxygen tube, I breathed in the fresh air of my homeland. There is nothing like arriving back home after being at war, no matter which state of mind or soundness of body you are in. It was as if God was whispering to me that everything was going to be alright.

With a police escort, we flew down the Beltway before arriving at Bethesda, Maryland. The notorious traffic did not hinder our convoy, as we moved onto the shoulder with the police sirens clearing the path. Pulling into the gate of Walter Reed National Military Medical Center (WRNMMC), I looked up at the vastness of the facilities. The hospital is a world all its own. And now for the next 13 months, it would be my home.

CHAPTER 9

THANK YOU FOR YOUR SERVICE
APRIL TO MAY 2012
BETHESDA, MARYLAND

The fourth floor of the Arrowhead Building on the WRNMMC campus was perhaps the busiest medical wing of the entire country in the
spring of 2012. Service Members of all branches wounded in the line of
duty during the Global War on Terror occupied the floor. Being a Naval
installation, the nursing staff naturally came from the Department of the
Navy, and they seemed to never have a break. Almost nobody residing in
a room was routine, most of the injuries being blast or gunshot related
from Afghanistan at this time.

I occupied a room to myself towards the middle of the wing with a
window overlooking the courtyard adjacent to the Arrowhead Building.
Apart from being back in the continental United States (CONUS), not
much was different. One of the first people I met was the liaison (LNO)
for the 82nd Airborne Division who was posted to the Warrior Transition Unit (WTU) at Walter Reed. SFC Grundy, another wounded
Paratrooper who had lost his eye in Iraq, ensured the transitions for
the division's wounded Paratroopers went smoothly and facilitated the
contact between the evacuated and the losing units. Accompanying SFC

Grundy every step of the way was his service dog, a beautiful black Labrador.

Within a day of arriving to Bethesda, I was reunited with Michelle. The look on her face when she saw me, though concerning, was reassuring. Out of anyone who could have walked away from a future with a disabled spouse and a lifetime of shouldering the needs that go along with it, Michelle never wavered.

The wedding was already pushed due to the deployment to June of 2013, so the MEDEVAC didn't change a thing. Despite the tubes sticking out of me from every limb, I embraced her and thanked her for being with me. She was only three weeks from graduating with her bachelor's in social work from Eastern Kentucky University. This girl, who should have been completing final essays and studying for finals week, was instead bedside with her fiancé. Eastern Kentucky University (EKU) has always been one of the most veteran friendly universities in the country. But to allow her to miss classes to travel to her future husband's side is one endearing aspect that I will always hold in high regard.

My parents quickly traveled to Walter Reed soon after I arrived. To my surprise as well, my brother Justin had been granted two weeks of rest and relaxation (R&R) from Afghanistan and he graciously came to my bedside to see me. I felt almost bad for them because for much of the time they were couped up in my small room or in a hotel when visitations for the day ended. Their company was needed however, as I hated being alone at this juncture and not knowing what my future entailed.

Despite the initial amputation at Kandahar Airfield and the follow-up at Landstuhl in Germany, more surgeries were needed to enable the residual limb to properly heal so that I could be fitted for prosthetics. My surgeon, Dr. Benjamin Potter, was the best in his field for orthopedic surgery and almost every wounded Service Member who cycled through Walter Reed knows him personally.

Within a week of my arrival, I underwent a major surgery that further amputated my left leg to about six inches below the knee. In this manner, my tibia and fibula were now cut in half and the skin folded underneath to form one scar. Pending any infections, I should be getting fitted for prosthetics once the swelling subsided, about two months from date of amputation.

Surgeries aside, this began a hard time for me in my life, a dependence on opioids, mainly Dilaudid. Now, prior to my injury, I never knew morphine. But when you're in pain, it is awesome, and you are extremely thankful that some chemist was brilliant enough to synthesize it many years ago. However, after the pain is gone, it continues to be awesome. See the issue here. You develop a dependency for what is basically clean heroin and mentally, you feel the need for narcotics to keep the pain, real or imaginary, away. For the first couple months, however, the pain was real enough, and with it a continuous supply of morphine. It was only after being released from the inpatient ward in July that things started to get hard. For now, it was very necessary for pain management.

One aspect of being treated at Walter Reed is the steady stream of visitors and well-wishers. From talented musicians, politicians, and celebrities; people from all walks of life, regardless of their political views, genuinely cared to come see us on the fourth floor. From my time there, Michelle and I met Jason Acuna (Wee man), Dave Matthews Band, and John Mayer. In addition to the celebrities, Generals, Admirals, and Command Sergeants Majors posted in the Capital Region regularly came to see us, always leaving a challenge coin as a small token of gratitude with me.

But perhaps the most important visitors I received during my stay were the President of Eastern Kentucky University, Dr. Doug Whitlock, and his wife Joanne. Doug, a former graduate of the EKU ROTC program and veteran, came to see me not long after he received word of my wounds from Afghanistan. His leadership and dedication to his students and the alumnus from his university is a testament to all. I am truly grateful for his leadership.

Not long after I arrived, a tour of the Pentagon was coordinated for the wounded Service Members. I remember the event extremely superficially but from what I recall, the mall that encompassed the Pentagon seemed foreboding to me, a young 2LT. A city within a city, the heart of the U.S. Department of Defense teemed with uniformed personnel from all branches in their dress uniforms. One of the event coordinators gave me a Crown Royal bag, to my amusement. This bag mystified me as the original contents were absent.

However, while being wheeled through the halls, a gauntlet of Field and Flag Grade Officers and CSMs met us and shook our hands. Coins from senior officers and NCOs began to fill that velvet bag quickly. It was neat, but perhaps a little too much for me, not even a week out from being blown up. I have yet to return to the Pentagon in any official or unofficial capacity and that one visit, while drugged, is my only lasting impression.

Exactly three weeks after arriving at Walter Reed, Michelle was scheduled to graduate on May 11, 2012. General David M. Rodriguez, Commander of U.S. Army Forces Command (FORSCOM) was to deliver the commencement speech and Oath of Office to ROTC's newest Officers. By sheer luck, I was cleared by Dr. Potter to attend the event to see Michelle on her special day.

The day prior to the event, I was driven to Andrews Air Force Base and loaded upon the small jet reserved for GEN Rodriguez. My mom, Vicki, who had been serving as my non-medical care attendant since Michelle had returned to finish her studies, accompanied me for the journey. We first flew to my home turf of Fort Bragg and landed at the adjacent Pope Air Force Base (now Pope Army Airfield) where we were to receive the FORSCOM Commander. GEN Rodriguez was incredibly gracious of me, and I deeply respect him for extending his hand to get me to Kentucky. We talked during the flight of my journey and how I ended up at Walter Reed. Touching down at Lexington Regional Airport, we were transported to Richmond.

I stayed at Michelle's parents' home in Winchester, Kentucky during my stay, my first time back in my home state since December of the previous year. It was great seeing them and all of Michelle's extended family. The next day, I donned my Army Service Uniform (ASU) with my dad's help and was driven to EKU. Looking at the pictures taken now, my ASUs were comically too big for me, I must have lost over 20 pounds in less than two weeks. I was wheeled to the front row of the cavernous Alumni Colosseum where the commencement and graduation ceremonies took place.

Although I don't recall who did, probably my dad, a large bouquet of flowers was set on my lap, for presenting to Michelle as she received her diploma from the University President, Dr. Doug Whitlock. During the National Anthem, I rose as best I could on a pair of crutches, notwithstanding the tangle of tubes still protruding from me. I had been heavily dosed with opioids that morning to get through it, but it took all the energy I had to stand. Despite the loss of a leg, I still believe I have no excuse to ever sit for the National Anthem.

Soon the names of the graduates from Michelle's college were called. Each student in their gown walked up to the stage to receive a blank rolled up piece of paper that symbolically served as their diploma. When her name was called and she walked onto the stage, I stood and slowly crutched over to her. Her smile was simply intoxicating, and she looked stunning. I was so proud of her for her accomplishment.

Handling her the flowers, I kissed her right there and then in front of the crowd of thousands. An eruption filled Alumni Colosseum. If this kiss stole the thunder from anybody else's graduation that day, I apologize. However, that moment holds dear to me to this very day, a powerful example of love and support that I owed to my future wife, for all the support and love that she had and continually has devoted to me.

The next day, I flew back to D.C. I couldn't hang out long in Kentucky. The risk of infection and my pain spiraling out of control was too great. The plane dropped GEN Rodriguez back off at Bragg prior to returning to Andrews Air Force Base. I thanked him again for facilitating my journey home to see my fiancée. It would not be the last time we would meet.

I settled back into my routine of healing, a mundane time in May that, although necessary, tested my restless soul. I needed out. Apart from the visitors and well-wishers, life on the fourth floor was anything but exciting. That was about to change however, as new forms of training that I had never encountered before were about to become my new focus: occupational and physical therapy.

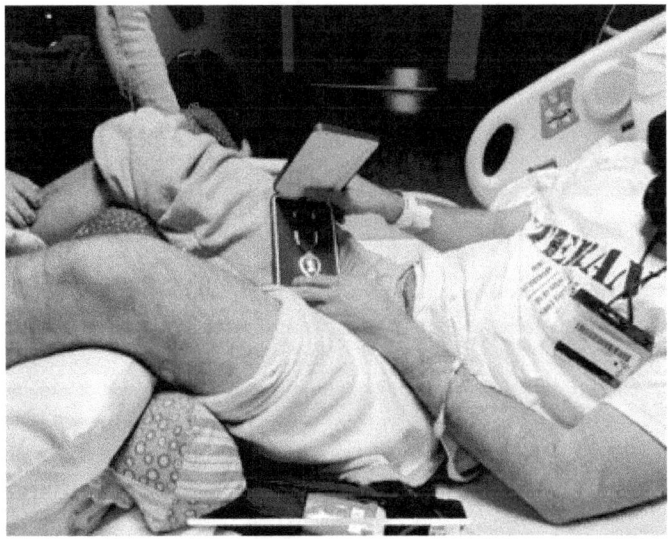

Awarded the Purple Heart at Walter Reed National Military Medical Center soon after being wounded. (APR, 2012)

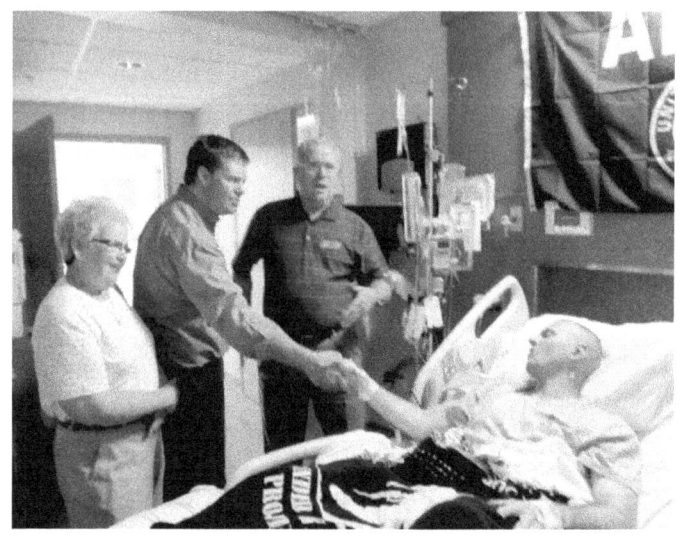

The President of Eastern Kentucky University Dr. Douglas Whitlock and his wife visiting Josh at Walter Reed National Military Medical Center soon after being wounded. (APR, 2012)

Josh flying home to Kentucky to see Michelle graduate at EKU three weeks after being wounded. (MAY, 2012)

CHAPTER 10
LEARNING TO LIVE AGAIN
JUNE 2012 TO JANUARY 2013
BETHESDA, MARYLAND

The Military Advanced Training Center (MATC) at Walter Reed is a state-of-the-art facility dedicated to rehabilitating wounded Service Members of all injuries. Encompassing a huge space with dedicated patient rooms, staff offices, fitness equipment, a therapy pool, and even a rock-climbing wall, there is absolutely no comparison from what I have witnessed. The staff, a medley of Navy, Army, and civilian medical care professionals, are some of the best therapists in their field concerning occupational, recreational, and physical therapy.

I remember being wheeled into the MATC on my first day of rehabilitation, a session of occupational therapy (OT) that focused on helping me adjust to completing activities of daily living such as getting dressed, grooming, transitioning, and bathing. My OT was a young professional from Coronado Island, California named Anne Marie Orr, OTD. Anne Marie has a cheerful demeanor who is incredibly passionate about her profession. Many wounded Service Members owe much to her for helping them adjust to their new lives with major limb loss or physical limitations.

As usual, I came into my first session of OT with an extremely cocky personality. I could do anything I wanted to and was not about to let a missing leg spoil my life's party. Although that is a great mindset to have, the truth is, I failed to initially grasp just how much I changed since the blast. The atrophy from being in a bed or wheelchair for over a month had caused severe weight loss, my balance was all but absent, and I struggled to perform simple tasks such as carrying objects, relegating my nurse or Michelle to fulfilling that role. Anne Marie was about to change that.

The aspects of OT that I do recall were that there was an emphasis on me regaining my balance, an important dynamic if I was to move on to physical therapy and learn to walk on a prosthetic. Now, the unique aspect of OT is that the therapist's treatment plan can comprise many different forms and techniques to help overcome a problem. Creativity is the name of the game with these trained professionals, creating unique treatment plans for every patient. On one session, Anne Marie had me balance on my right leg about six inches away from a waist high table while disassembling and assembling a non-firing M4 Carbine. Prior to losing my leg, this was a trivial matter that I had executed hundreds of times in the years before. Now however, I struggled to perform the basic steps because I had to shift my focus to not touching the table or falling on my fourth point of contact. It took many attempts to solve this riddle, but once I had it down, she threw another test my way.

Legos are some of the best forms of entertainment out there. With imagination and enough pieces, you can build just about anything. I was given a box of Legos; I believe a Star Wars set of a spaceship. Anne Marie directed me to put it together just like the M4 problem. Just opening the sealed box and putting the pieces on the table was a chore. Now I had to focus on balancing, reading IKEA like instructions, finding the corresponding piece, and assembling a science fiction space craft. I could have spent hours trying to put this together. But my sessions lasted typically for an hour due to fatigue and follow-on appointments with wound care or pain management. So, over the course of several sessions, a product was constructed to perfection.

Alternating between balance tests was mobility training. I practiced crutching up and down stairs under her supervision, pistol squats to build my right leg, and massages of the residual limb to combat phantom pain, a sensation where an amputee still feels as though they have their missing limb. It can range from the feeling of being tickled or a white-hot nail being driven into the absent limb. I still get it on occasion, usually when I am driving or standing in one place for too long. On one session, she had a Nintendo Wii, and we did karaoke to Elvis Presley while I balanced. What a sight that must have been. I hope to God to this day she does not have that video.

And throughout all my sessions, there was Michelle at my side. She had traveled to Walter Reed to be my Non-Medical Attendant (NMA) during my stay. She wheeled me to the MATC every morning and accompanied me to every appointment. Never has a man ever had such devotion from a woman who was not yet his wife. And never did she ever complain about it. She gave up her free summer with her friends, family, and starting her career just to help me rehabilitate. To this day, she has always been my greatest champion. A benefit of being an NMA is that you get paid for it.

As the sessions of OT passed, my balance slowly recovered. I could confidently crutch upstairs without issue, transfer from a wheelchair to a shower or vehicle, and could be trusted to take care of my private needs without fear of falling. Now, one thing is for sure, I became an all-star at hopping. If there was a hopscotch competition, I would take the gold medal. However, a lesson for all amputees out there, DO NOT HOP. It is terrible on your knee, hips, and spine. Don't do it. The therapists at Walter Reed will scowl at you and scold you brutally if you do it in front of them. Use your wheelchair, crutches, or prosthetic. You'll thank me later.

After a month of occupational therapy, I was reassessed by Anne

Marie and the determination was made for me to start physical therapy to build up my stamina and strength. The healing process was progressing, and I was soon to be moved out of the fourth floor and into the Warrior Transition Unit Barracks at Building 62. Within a couple weeks, I could be evaluated for fitment of my first prosthetic leg, the ultimate goal in being able to return to duty.

Physical Therapy is executed adjacent to the OTs in the MATC. Throughout my tenure at Walter Reed, I had two physical therapists (PT), Dr. Bo Bergeron, DPT and Kelly McGaughey, DPT. If OT was fun, PT was exhausting. Bo, the Chief of Physical Therapy for the MATC, was my first PT. Immediately one could see that she was highly respected in the tight knit group at the MATC. Her approach to me was very direct, to the point, and showed that she was all about business.

My muscles atrophied while I was on the fourth floor in the in-patient ward. I needed to rebuild the strength in my lower body to be able to walk again. For an hour, five days a week, Bo had me stretching and doing strength conditioning workouts. From leg curl machines, Pilates, to scooting around the MATC on a wheeled chair with weights on my right calf, I was beginning to rebuild my right leg's strength. My lower body was not the only targeted group, as rope pulls, hand crank machines, and dumbbells were employed to facilitate muscle growth in my arms, chest, shoulders, and back. Although the MATC is not a Gold's Gym, Bo employed many techniques to assist myself in regaining my strength.

The MATC was one of the few times where I could see my fellow comrades, who also were embarking on their paths to heal. Every type of injury imaginable was being treated by these therapists. The dark humor in this facility was intense, as the Navy, Marine, and Army patients derided each other. Those with the worse injuries often threw the harshest jokes at us. To combat them, some of us would instigate the most brutal jokes of our own. Red Cross volunteers and interns were sometimes horrified, but for the most part, everyone knew it was our way of getting through the healing process.

In June, I attended my first recreational therapeutic event. Harvey Naranjo, COTA was the go-to guy at Walter Reed for any on or off-site activities for the patients. From kayaking and swimming in the base Olympic sized pool to working out at the Under Armor Headquarters amazing fitness facility in Baltimore, Harvey had something for everyone. One of the jolliest dudes I have ever met, Harvey can only be described as a one-of-a-kind soul whose tireless work ethic ensured smiles all around from the patients and their families.

The event I attended was a 5K race sponsored by the Achilles Corporation and U-Haul that took place in Central Park in New York City. As I was still relegated to crutches, I would be hand biking during the race. In addition to the race, there was a banquet by the event's hosts, a jet boat tour of New York Harbor, and we were to attend a New York Yankees and Mets baseball game. Michelle and I had never been to New York City before and we relished the opportunity to see the biggest city in our country and one of the biggest in the world.

The commercial bus pulled up next to Building 62 where the wounded Service Members who were no longer inpatient resided. Wheelchairs, crutches, suitcases, and prosthetics of all varieties were being loaded as the patients, their non-medical attendants, and various accompanying family members boarded. Harvey was the last to load, with his ever-present Chocolate Labrador, Deuce. The journey took us from the Beltway to I-95 and straight to New York City, passing Baltimore and Philadelphia along the way. Apart from my short visit to Kentucky to observe Michelle's graduation, this was my first road trip since deploying in January.

After four hours, I could see the New York City skyline to my left. I just couldn't fathom the magnitude of such a vast urban center. As we journeyed closer, the skyscrapers seemed to reach for the clouds. It was truly a marvel, especially for a military brat who grew up in communities where the highest structures were the 250-foot parachute towers at Fort Benning, Georgia. A police escort took us to our final location in Manhattan. Each patient was billeted in a hotel of some type in the immediate vicinity. After disembarking and given instructions on where to meet, Michelle and I moved by foot or crutch to our hotel a couple blocks away. The road was paved with trees and flower plots stood on every stoop. This was a nicer part of the city I observed, no doubt with addresses here costing well into the millions of dollars for the bragging rights of ownership.

The hotel was built next to some townhouses. At first glance, there was nothing to make it stand out, apart from a small sign. Crutching up the stairs, I observed the creole artwork and decorations for the New Orleans theme; very eccentric but cool, nonetheless. Despite being exhausted, our evening was packed with events. We had a visit to a NYC fire station to meet some of the fire fighters who responded to the events of September 11, 2001 at the World Trade Center. Then we had a banquet hosted by Achilles and U-Haul for the run and the wounded Service Members.

The bus took us to a fire house in the city that looked older than the skyscrapers surrounding it. To our surprise, James Gandolfini himself was at the station to meet us. Humble and with the personality to match his Sopranos character, it was a cool experience to talk with him about the city and he was very interested in our stories as well. The first responders at the station explained how they responded to the terrorist attacks and solemnly showed us a memorial to those of their station who gave their lives. Although the attacks had happened almost twelve years prior, you could tell it was still on their minds. It was eye-opening to see that Service Members in the U.S. Military were not the only ones who had bled for their country when under attack.

After the fire house tour, the group executed several excursions throughout the Manhattan area over the next day. We toured the World Trade Center complex, to include seeing the Freedom Tower in its final stages of completion. Perhaps most interesting was a jet boat scenic ride throughout the Upper Bay where we viewed the Statue of Liberty and had a water cannon salute from the fire fighting ships. The tempo didn't lighten up that evening. Achilles and U-Haul hosted a banquet for the group which was immediately followed by a trip to Citi Field to watch the NY Mets take on the NY Yankees. After the game, each of us received a signed baseball from the NY Mets. We finally arrived back to our little hotel late that evening, utterly exhausted from that day's activities. Tomorrow was the 5K throughout Central Park, followed by the journey back to Walter Reed.

I had never hand biked before. I was representing a fallen Ranger from the 3rd Battalion, 75th Ranger Regiment. I had no idea what was in store for me as I crutched off the bus and viewed the dozens of hand bikes. I had to make a good showing in this race, especially since the mother and sister of this hero were here to support me. Some of the other wounded Service Members with amputated limbs were running on their prosthetics, an aspect which made me jealous as I had always been a runner. My time would come though.

Although I was complete in less than twenty minutes, the race felt like an eternity. I had grossly underestimated the strength required for hand-biking. To spice up the situation, the race course seemed mostly up hill. By the time I crossed the finish line, my forearms, biceps, and shoulders were shot. I had not let up once on those darn cranks, yet I finished nowhere near first place. Although I didn't quit and gave my all, I still felt that I had let that family down whose son I was racing for. Nonetheless, a giant hug and smiles greeted me at the finish line where the two woman embraced me as one of their own. Michelle too was there, beaming as she always does when I'm doing something I enjoy.

The ride back was filled with sleeping Service Members, crying babies, and spontaneous chatter from the families. Reflecting on that trip, Michelle and I saw so many amazing facets of our country in just one city. The people of New York City are some of the best around and their resiliency had not wavered in the eleven years after September 11, 2001. In addition, a newfound outlook on life was beginning to take hold. Having one leg was not going to keep me from living my life and loving myself and others. My therapists devoted so much time and patience to me and I was damned if I was going to waste the hours of rehabilitation on complacency. From that point forward, nothing was going to slow me down and achieving my goal of returning to duty with no limitations.

On the day I was supposed to be cleared for prosthetics, two surprises awaited me. First, my high school friends, Adam Ward and Scott Stafford, came up to visit me after being appointed as newly commissioned Officers in the U.S. Army. The second surprise was not as uplifting. A fever had struck me the previous night and I began shaking uncontrollably and shivering. After a bout of PT, Dr. Potter came over to clear me. Upon closer inspection, I was immediately sent back to the 4th Floor. He noticed a thick yellow discharge oozing from my leg and the swelling was too high to indicate anything other than a serious infection.

Within hours, with my two friends and Michelle hanging out in the room, I was back in surgery. I would have to wait another month at least for the wound to heal and the swelling to subside to the point where a prosthetic could be fitted. As the intravenous line was inserted and the pain killers administered, I felt as though I was back to square one, which I kind of was for the healing process.

But thankfully, the OT and PT would enable me to get through this much quicker once the medications wore off and I was back on crutches. For 30 more days, I healed and waited, until finally being cleared to learn to walk again for the second time in my life.

By late June, I was fitted for my first prosthetics in the on-campus factory in the America Building. The sockets were made of a plexiglass type material and seemed to weigh a ton. Measurements were done to ensure the leg was calibrated perfectly to ensure my hips and back were not out of line. My prosthetist, Roger, was incredibly patient with me as I just wanted to take off with it and ditch the crutches. The prosthetic was also incredibly uncomfortable at first to put my full weight on. My PT, Kelly, knew just what exercises I needed to do to get used to walking on that leg.

Initially, I was confined to only walking in the MATC until Kelly was comfortable with me walking outside of the medical facility. The test to take the leg home was simple enough, lay down flat on the ground and get up to the standing position without assistance. As expected, I aced it and walked out of the MATC that day, confident that my crutching days were over.

Little did I realize though how difficult it was to adapt to a prosthetic. The heat from the silicone liner made sweating an issue and sores occasionally rose along my residual limb from the rubbing. On short walks, I could muscle through the cramps and pains of the leg, but on longer walks I needed the support of a cane. It was by no means a quick turnaround and I now understood Kelly's advice on taking it easy.

However, the more I was active on the leg, the more the swelling subsided and thus became less painful. Before long, Michelle and I were taking walks throughout the campus. On the recreational therapy trips, I had a new feeling of freedom that compared to nothing I had experienced before. Everyone takes walking for granted, the same as breathing and blinking your eyes, until you cannot do so anymore.

I was getting stronger too, physical therapy being transitioned to strength and conditioning training at the fitness center at Walter Reed. I was on track to train to return to duty.

The previous trip to Manhattan was the first of many trips that Harvey and the team took us on. Throughout the remainder of 2012 and well into 2013, I undertook many recreational therapy programs from biking, horseback riding, snowboarding, hiking, and all other types of activities to rebuild my confidence and adapt to the challenges thrown at me. While the OT and PT treatment plans got me back on my feet, the extracurricular activities truly helped me find my passions for the great outdoors.

I visited New York City again that summer where I stayed at the beach home of Tom and Gretchen Fox at Breezy Point near the Rockaways. I learned to surf and even sailed through the bay on just one leg. Tom and Gretchen are two of the best souls I had ever met, and their genuine love and concern for me and Michelle forever endeared them to my heart, and we stay in touch to this day.

Towards August 2012, Michelle and I moved out of the outpatient barracks of Building 62 and into a one-bedroom apartment in Rockville, Maryland, only a few miles north of Bethesda. This was to be the first residence that Michelle and I lived together in. And to add to our small family, we adopted a Boykin Spaniel puppy named Aussie.

As we had no furniture, IKEA became our go-to for filling out our dwelling. And let me inform you this, please buy an actual tool set and drill prior to assembling those pieces of furniture. You will not only save a lot of time and frustration but will not have to deal with the accompanying wrenches that aren't worth a box of used chewing gum.

Apart from living off Walter Reed's campus, my routine remained the same. Physical therapy in the morning at 0800, regularly scheduled appointments until around 1300 hours, and then the remainder of the afternoon typically free. Most of Michelle's and my free time consisted of working out, traveling around the Capital Region, or shopping in one of the many malls in the area.

Even though the trips with the MATC crew had kept my motivation high, I yearned for my family and friends back home in Kentucky. The opportunity came in late August when Scott Stafford invited me to go fishing with him and his dad at his cabin in Minnesota. Just prior to that, I made one of the most important decisions of my life. I flushed down the remainder of the prescription narcotics I had on the day we traveled north from Elizabethtown, Kentucky. I was going to get clean and break free from the mental dependence I had developed on painkillers.

The week in Minnesota proved to be one of the best times of my life.

Waking up early, we drove out to a different lake every day and fished off Scott's dad's boat until dusk. Never had I caught such monster bass, one of which is mounted on my wall today. We took my Dodge that my father had brought up from Fort Bragg to Rockville, so I had some wheels. In one instance of insanity, we went off-roading adjacent to his father's cabin, which coincidently ran through someone's soybean crop. While we had the time of our lives, I later learned no trespassing signs were installed on the neighbor's property. Sorry about that, Scott. When I returned to DC, I was clean and more importantly, ready for the next chapter of my healing journey, the fight to return to duty.

Task Force Fury was scheduled to redeploy from Afghanistan back to Fort Bragg that Fall. The 82nd Airborne Liaison, SFC Grundy, ensured the score of wounded Paratroopers made it to Green Ramp at Pope Army Airfield to see their respective unit's return. Standing alongside another wounded Paratrooper, SSG Erik Myers who had lost both of his legs from the waist down, we were asked to march into the hanger with our comrades after their disembarkation from the aircraft. It was surreal for me to see my brothers and sisters in arms with whom I served on my first combat tour.

Although many questions were asked about how I had been, I wanted to shy away from it as much as possible. To me, this wasn't my moment but theirs. I already had my homecoming, albeit a little unorthodox. These Paratroopers had completed their mission in Afghanistan while I recovered. They deserved their time in the spotlight.

The prosthetic I walked into the hanger with was awkward and keeping the pace and in step with the formation was difficult for me. SSG Myers' wife, Laura, pushed him in his wheelchair so I could not openly complain. Nonetheless, I didn't want my first impression as an amputee in front of my unit to be one of pity. I had to remain strong and support my comrades.

Green Ramp teemed with friends and family of the returning Paratroopers from 2-508th PIR. Senior leaders were there as well, Brigadier

General Charlie Flynn, now the Deputy Commander, 82nd Airborne Division and the FORSCOM Commander himself, General David Rodriguez, were there in attendance. General Rodriguez, who had previously flown me to Kentucky to see Michelle graduate inquired as to how I was doing and seemed genuinely concerned about my progress. After saying farewell once again to the Paratroopers I served with, I returned to DC, not to see my comrades again for quite some time.

A week later, I flew to Crested Butte, Colorado for a week with our occupational therapists for rock climbing, hiking, mountain biking and kayaking, immediately after seeing my comrades return from war. My first time in Colorado, I truly fell in love with the jagged peaks of the Rockies and the lifestyle of the populace. On specially modified hand-bikes, the group mountain biked on trails down Crested Butte and through the town.

One day, the option for the group was as follows, either kayak on a local lake or hike Crested Butte. Selfishly, I wanted to hike to the peak of the 12,162 ft promontory while most chose to kayak. A local guide took me and one other hiker from the base to the top. In a couple of hours and slightly suffering from altitude sickness due to a lack of acclimation, I summited. I had climbed a legit mountain only months after losing my leg.

Colorado and the Rockies had my heart and I endeavored to return. Thus, in December Harvey offered me an opportunity to go back for a week of snowboarding at The Hartford Ski Spectacular. Hosted in Breckenridge, the Hartford Ski Spectacular brings in disabled Service Members and civilians alike for a week of adaptive skiing and snowboarding. Put on by The Hartford and numerous other organizations, it was without a doubt one of the most influential events that I attended. Being an avid snowboarder, I signed up without hesitation; my only regret being that Michelle could not accompany me.

Snowboarding on one leg came natural to me. I brought my own board and needed little tweaking to accommodate a prosthetic foot. I felt a new form of liberation as I flew down the slopes on my Forum Destroyer snowboard. The hills I grew up on in Kentucky and Indiana dwarfed in comparison to the slopes of Breck. Apart from the opening and closing ceremonies banquets, I was free to wander the picturesque town of Breckenridge in the evenings. All dolled up for Christmas, it was a beautiful sight at night. Michelle would have died and gone to heaven with the snow, lights, and the shops. I spent most my evenings dining in the town at the local shops and sampling the beers at the local breweries. By the end of the week, I had made some awesome friends and had improved so much on my snowboard that I was invited on another trip in January, a weeklong excursion to Vail, Colorado where my fiancé could participate.

If Breckenridge was beautiful, Vail proved to be heaven on earth. The Rockies, when standing on any peak of the slopes, looked exactly like the Coors Light can, and just as cold and refreshing. Michelle, who had never skied or snowboarded before, soldiered on throughout the day learning how to navigate the slopes by dedicated instructors. Not wanting to regulate myself to the basic trails she was learning on, an instructor named Mike took me all over the resort's hills. Ranging from moderate to difficult, I explored the bowls and terrain parks of Vail with Mike who had nothing but a grin the entire time. Mike had the personality of a laid-back surfer who loved to travel and meet new people. He showed me techniques in the terrain park and on powder that greatly improved my abilities and confidence on the snow.

During the evenings, the group was hosted at different diners, clubs, and bars. Unlimited food and opened bars to all the wounded Service Members. It was as if the entire town of Vail had thrown open the doors to us. Out of all the trips Michelle and I attended, I grew so close to her while in Vail, pleased that I had found the love of my life who enjoyed the same activities as me. More importantly, she also witnessed how much support we had from our country. As a future spouse of a wounded Soldier, she was never going to be alone or without resources. People all over the country were bending over backwards to support us in any way they could as a patriotic expression of their gratitude or simple gesture of their generosity.

Our final trip prior to my departure from D.C. occurred shortly after Vail. Soldier-Ride was a wounded Service Member bicycle ride from Miami to Key West. Michelle was also able to accompany me as a non-medical attendant. This was incredibly beneficial as she assisted me during my first ride on a bicycle since losing my leg as opposed to a hand bike. Harvey, as always, led the charge with organizing and leading the group, ensuring strict adherence to timelines.

The first leg of the trip, biking through Miami had us stopping at the Miami Marlins baseball field for a meet and greet followed by dinner at a venue called the Stone Crab. Easily now one of my favorite restaurants, I proved incapable of stopping myself from eating too much stone crab, hash browns, and key lime pie. Afterwards, we proceeded the next day down Highway One towards Key West. Biking through the islands that encompass the Florida Keys, we stopped for food and at one point, to swim with dolphins.

Once arriving in Key West, we were free to our own devices. Touring the Ernest Hemingway House, seeing live music, drag comedy, and exploring the Naval Station on the Island, it was quite the contrast from the winter activities I had been performing only a week prior. We even won a free weekend down at an ocean side cottage in town. Once again, these experiences brought me closer to Michelle while empowering me. I no longer was shy about my amputation but proud of it. I could do anything I wanted to, and I always strove to prove that physical limitations were just a rumor for me, they simply did not exist.

Recreational Therapy was incredibly valuable, not only to my physical recuperation, but to my mental wellbeing. I was not only taking back my life through the experiences that gave me confidence and boosted my self-esteem, but also discovering a new love for outdoor recreational activities that were put on hold throughout college and military training.

Whenever I meet someone who has suffered tremendously, I always inquire now as to if they participate in any recreational hobbies. If they are unsure as to if and how they could do something they used to love despite their abilities, I try to point out the examples that I have witnessed as a testament that with today's technology in medicine and adaptive equipment, almost anything is possible.

Despite the barrage of activities, I strove to participate in, life was far from perfect for me. On the outside, it seemed like my recovery was flourishing. Inside, however, I was screaming. Depression took it's toll immensely during those months; especially when I was in the outpatient facility of Building 62. The pain killers I took at the beginning of my stay now failed to numb my mind. When I finally quit the medications, it was as if a new monster moved in.

Ask Michelle and she will inform you that our situation was tenuous. I had fits of anger and at times rage. I became angry at the world and everyone in it. To my shame, I took it out on Michelle many times. Not because it was her fault, but due to the fact she was the only person I could confide in. She was with me day in and day out and witnessed a part of her fiancé firsthand that she had never encountered.

I recall one incident while I was in a bout of anger when Michelle tried to confront me. My response was a couple of poorly aimed ceramic dishes projected towards her. I was angry because the staff were adamant that I take my medications. At one point, I was under so many opioids that I physically felt ill and couldn't function. When she tried to tell me that they were ordering me to take my prescriptions as dictated, I lost all control.

She called one of the staff orderlies in the Warrior Transition Unit who came to my room. The NCO, a SSG, saw the broken plates and asked Michelle if she needed me to be separated. Michelle, as loyal as ever, declined and stood up for me. She could've packed her bags and departed back to Kentucky, but she stayed by my side and even defended me, explaining that the medications were too much. Only after additional intervention from the pain management team were my doses lowered.

Physical Therapy at Walter Reed, (JUL 2012)

First handbike race at New York City, (JUN, 2012)

CHAPTER 11
AGAINST THE ODDS
SEPTEMBER 2012 TO MAY 2013
BETHESDA, MARYLAND

My first test on how my rehabilitation was progressing occurred that fall. In October the Army Ten Miler race occurs every year. With a route beginning and ending at the Pentagon and winding through the National Mall and some of the historic landmarks of Washington D.C., it is one of the most famous races in the United States. Since running ten miles was still out of the question, I decided to conquer this route while marching with a 35 pound rucksack. Using the Ranger School standard of 15 minutes a mile with the prescribed weight on my back, I wanted to prove that I could still move for long distances with weight on my back, a prerequisite for any infantryman.

On a chilly Sunday morning, a commercial bus of rehabilitating Service Members and a U-Haul truck of hand bikes and prosthetics staged near the starting line. Per tradition, wounded Service Members started the race with a head start prior to the other competitors. Harvey, as usual, was organizing the movement of our group to the starting line, he himself having collected dozens of Army Ten Miler Finisher coins over the years. Michelle was going to walk with me in this race. My old

ROTC buddy, 2LT Caleb Wood was also marching with me, having traveled to D.C. from Fort Bragg after his own tour in Afghanistan. I was wearing an Army Ten Mile tee shirt and black shorts, commonly known as silkies or Ranger Panties.

Upon the initiation of the race, a mass of hand bikes took off from the start, followed immediately by runners utilizing prosthetic blades in lieu of running shoes. The three of us took off amidst the runners. I was feeling good about this race; the atmosphere energized me. Winning for me at this point wasn't about where I placed but rather knowing I could still do what was required of me to lead Soldiers. Caleb kept pace on his watch, and we were on track to finish in two hours and thirty minutes if I kept up. This was Michelle's longest walk as well, a small accomplishment for her and I am proud to this day that she endeavored to complete the race with me.

Along the route, many competitors encouraged us, to include members of my own unit. I felt good as I marched forward, the leg slightly rubbing but nothing that alarmed me. After the eight-mile marker, I finally began feeling the cramps associated with wearing a socket and the rubbing induced by the friction between the skin and socket sleeve. Finally, after two and a half hours, we crossed the finish line and ended my second Army Ten Miler, the first having been executed as a Cadet back at EKU. My old Professor of Military Science, LTC Rich Livingston, was even there at the finish line to congratulate us. It was an awesome feeling and a boost of morale to know I still could perform to the standard.

The next hurdle was passing the Army Physical Fitness Test (APFT), a necessity if I had any hope of remaining on active duty. Although I could have been exempt from this bi-annual requirement due to physical limitations, I did not want that treatment. I didn't want anybody to treat me differently due to my wounds, so I trained to be able to accomplish every task asked of a Soldier. Kelly worked with me in the MATC to build up my strength. Some days, I jogged around the Walter Reed campus to rebuild my stamina. Every day had a different pushup or sit-up workout for me to complete.

I was growing more comfortable and stronger in my leg, and this was still a basic prosthetic, long before I had my current Flex-Cheetah and Flex-Explorer limbs that allow me to move without any indicators of an artificial limb, gracefully and naturally. Prior to going to Ranger School, I was in the best shape of my life and could max the pushups, sit-ups, and two mile run with ease. Now the run was my hardest event to max.

2013

On a cold February day, one of the interns met me at the track at a local high school where I was administered the APFT. I maxed my pushups and sit-ups and achieved a good run time. When all the event's points were added up, I had achieved a score of 283 out of 300 points, the lowest APFT of my life. Nonetheless, I was still in better shape than most of the U.S. Army with one leg. With that obstacle overcome, the final battle was convincing the Physical Evaluation Board (PEB) that I still had what it took to remain on Active-Duty Status.

The Physical Evaluation Board of the Integrated Disability Evaluation System (IDES) took up much of the winter and early spring at Walter Reed. A process that determines whether a Soldier can remain on Active-Duty Status or should be medical retired or discharged, the PEB was my last obstacle to return on Active Status.

Unfortunately, it was rare to see amputees on Active Status. Most had respectfully retired from the force or reclassified into different Military Occupational Status' (MOS) that would better accommodate their needs. I had to convince a panel of people who knew nothing about me that I still could serve without being a hindrance or liability to the unit I was assigned to. Countless interviews between a liaison officer at Walter Reed and the Veterans Affairs, as well as medical appointments with my primary health care providers were needed for the board to have enough evidence to decide. It was a slow process, full of frustration, and at times it seemed as though the personnel whose role was to assist me throughout this process just wanted me to take the easy way out, medically retire.

But there was no way in hell I was going to retire at the age of 24. I had worked my ass off after over five years of training and education. After qualifying as an Infantry Officer, Paratrooper, and Ranger, I was not about to change my occupation and branch to a support role. I proved during the Army Ten Miler that I could ruck, and I completed an ACFT with a decent score. I was in shape and had no significant mental or physical issues lingering from my initial blast and the follow up surgeries. I just wanted to go back to the 2-508th PIR at Fort Bragg and resume my young career.

For months, I awaited the results of the board. I felt complacent at this point, the recreational therapy no longer challenging me, and I had long since completed my physical therapy. I worked out every day at the modest fitness center in my apartment in Rockville. Michelle and I waited anxiously. I took up as much of my free time improving myself professionally by interviewing and achieving internships with federal agencies and obtaining security clearances. I wanted there to be no doubt that this officer was still needed in the force. If anything, I wanted to have so much time and money invested in me that it would be a huge loss to let me go. Then I received my wish.

On March 19, 2013, the PEB approved my wish to return to active-duty status with no limitations. As far as the government was concerned, I had no disability rating, and I had a clear bill of health. Michelle and I were ecstatic to finally leave and return to Fort Bragg. Michelle was full speed ahead in planning for the wedding that summer on June 22nd and was even more excited about looking for homes in which we would reside as husband and wife.

Unfortunately, it would take another two months before I received any orders to PCS back to Bragg from the Warrior Transition Unit.

Annoyed and impatient, I did what I thought was right by contacting my Senator, Mitch McConnell and my old Brigade and Battalion Commander COL Brian Mennes and LTC Guy Jones. They wasted no time in getting me back into the fight and utilized the resources available to them to retain me.

That spring, I was invited to what would be my final event in D.C. A golf scramble in Rockville, Maryland was being put on in memory of Mrs. Melanie D. Strudler who had recently succumbed to cancer. She had been a volunteer at the MATC for many years. Her daughter Erica, and friend Sheryl had overcome the insurmountable task of putting this venue together from scratch. In addition to honoring her mother, Erica wanted to honor the Service Members as well and extended invitations to Walter Reed. Wanting to get back into golf prior to returning to Fort Bragg, I signed up.

On a beautiful May morning, I played eighteen holes with some of the most friendly and genuine people I had ever met. The organization Erica had built, Miracles for Melanie, is indeed one of the best charitable organizations I have come across. What was once two young girls trying to figure out how to put on a golf scramble has blossomed into something much greater than I bet they could have imagined. You can tell by the passion how much love for Melanie there is. I can only hope that I can live a life where my legacy is remembered with so much fondness.

After the game, I talked to Erica and Sheryl about their cause and how they came about putting it all together. I was incredibly impressed and promised to come to similar events afterwards when my schedule allowed. Unfortunately, something else was in store for me for the future that hindered me from fulfilling that promise.

At the beginning of May, I finally had a set of orders. I wasted no time in clearing Walter Reed. I made appointments for movers to pack up my stuff to ship 330 miles south to Fayetteville. Michelle even found a nice house to rent in the meantime where our furniture would go.

Saying my goodbyes to Anne Marie, Kelly, Harvey, Roger, and all the medical professionals who had enabled me to succeed was the hardest part. But at the same time, I could see how pleased they were that I was going back to my old unit and resuming my career. It was such a rarity for Service Members with major loss of limbs to return to duty. I was one of those good news stories.

Out of the amputees at Walter Reed who were wounded in 2012, I was a lucky one. To my medical care providers, I was their victory by returning to duty, because of all of the hard work and effort they dedicated to me. Had the blast been bigger, that probably would not have been the case. I was blessed. I was fortunate. But most importantly, I was surrounded by people who cared so much about Michelle and me. Friends and acquaintances who went to work every day trying to assist Service Members like me in recovering had a good news story. I was blessed with this second chance and was not going to squander it for anything.

If I could provide any wisdom for wounded Service Members who want to return to the fight or remain on operational status, my biggest advice is to be patient, articulate clearly to your health care professionals what your goals are, and be realistic. Know yourself and what your occupation demands of you, and work closely with your team to devise a plan and timeline to achieve your goals. Be ready to accept setbacks, but never accept defeat. Motivation fades but discipline endures so make that routine and stick with it. You are your biggest enemy, don't let self-pity or doubt overcome what you are truly capable of achieving.

Patience, or lack thereof for me, defines my 13 months at Walter Reed. I desperately wanted to return to Afghanistan only months after being wounded, but knew that wasn't going to happen. I wanted to go for a run immediately after trying that prosthetic on, but it wasn't going to happen. There are processes and procedures in place for the Service Members unlucky enough to be sent to Walter Reed for combat wounds and injuries. And you are not an exception to the rule. While waiting for your body to heal and recuperate, fill your schedule with as many activities as possible. Attend the recreational therapy trips and activities that you are invited on or see posted on the bulletin boards in the MATC. Even if you never biked, snowboarded, kayaked, etc., get on the list and give it a go. You never know what you are capable of until you try it. Even more important, you never know what you might discover as your new passion.

Your nurses, doctors, surgeons, physical therapists, occupational therapists, recreational therapists, prosthetists, Warrior Transition Center cadre, and Physical Evaluation Board Liaison Officers should know without a doubt what your intentions are soon after arriving at your medical installation for treatment. They will have their plans for you regarding surgeries and therapy but assist them in knowing what you want as your end state or goal.

Do you plan on retiring and staying at home? Transitioning from the military to learn a new trade or going back to school? Or going back to your old unit and retaining operational status? There are many paths you can take. None are wrong and none better than the others.

However, communicating your goals may help your team of supporters assist you and modify treatment and therapy plans to help you accomplish those goals. Nobody wants to see their patients fail or relapse. These medical care providers and military and civilian professionals' endeavor to help you live your life to the fullest after the most traumatizing and severest of injuries. Be open and transparent. Allow them to help you achieve your goals. You've just had a setback. Let them show you how to overcome the next ones.

Finally, be realistic. I will no doubt get some flak for this, but I cannot stress this enough. It is incredibly hard to return to Active-Duty Status with limb loss. But it is possible. After the scope of your injuries is known, what other enduring problems do you have? Can you overcome them? What equipment do you need to adapt to the challenges, and will it hold up to the rigors of life in austere and non-permissive environments? What is the availability and accessibility of new equipment when your hardware fails? What is the mission of your unit and what tasks must they train on regularly to be ready to accomplish them? Finally, will you be an enabler or liability to your organization in the event your unit is deployed into combat?

These questions must be answered and researched thoroughly throughout the entire process should you choose to remain on operational status. There is no shame in reclassifying your MOS or taking a role in sustainment to support the team pulling the trigger. You are the only person who knows your capabilities and should endeavor every day to push yourself to the max to find your threshold of failure.

Once you discover what you need to improve on or cannot accomplish, develop a plan, or research the waivers out there. But never give up. Have a Plan B. In the event the PEB findings are determined to not be in your favor, what is your backup that will suit your needs and help you continue doing what you love to do?

Be realistic about what your goals are but ensure they are never too small or easy to overcome. Push yourself during therapy. Inspire your OT and PT to adjust your treatment plans so that future appointments are absolutely hell on you. Take your adaptive hardware and make it not only a part of your body but a part of your soul. There is no going back after a limb is amputated or eyesight vanishes.

However, in this modern age, there are resources available to you in the United States by virtue of your service that will ensure you will not live a complacent lonely existence. Just ask SSG Travis Mills, who created his own retreat and foundation after losing both arms and both legs to an IED while patrolling with Task Force 2-Fury in 2012. If he can accept his future, overcome his wounds, and flourish, then I think anybody could, given the circumstances and opportunities they have.

On May 14th, 2012, after exactly 13 months in recovery and rehabilitation in the nation's premier medical treatment facility for Service Members; Michelle, Aussie, and I traveled south on I-95 for the five-hour journey home. We followed behind our moving van in my truck and her Audi. We were excited about our future together. It was only one month until we were getting hitched in her hometown of Winchester, Kentucky.

I was set to be a Paratrooper again in the 82nd Airborne Division and lead Paratroopers during the Global War on Terror. Our adventures in D.C. at a conclusion, a new chapter in our lives was about to begin.

Josh, Michelle, and his friend 2LT Caleb Wood rucking the Army Ten Miler six months from being wounded. (OCT 2012)

CHAPTER 12
TRANSITIONS
JUNE TO JULY 2013
FORT BRAGG, NORTH CAROLINA

Little had changed at Fort Bragg from the sixteen months since I had departed for Afghanistan. With enthusiasm, Michelle and I pulled off I-95 and onto the bypass towards Fayetteville. We had found a house in the southern part of Fayetteville, roughly halfway between the city center and the town of Hope Mills, in a subdivision. Based off the photos from the rental agency, the home we chose to make our own was in a new subdivision and had all the amenities a couple and a puppy could need.

The enthusiasm Michelle had quickly turned to gloom as we pulled into the neighborhood. For lack of a dedicated reconnaissance, her eyes quickly widened as the area teemed with signs of poverty and crime. I believe she almost cried before we even reached the home at the back of the neighborhood, calling me and apologizing for picking this location (I was in the truck behind her). I had hope though, and had seen and lived in much worse, having grown up in the military housing of the 90's that can best be described as spartan accommodations at best.

Pulling up to the house, she wasted no time inspecting the interior.

Satisfied that the pick she chose wasn't a complete loss, notwithstanding location, we waited patiently for the moving van to arrive later that day. Since we came from a small one-bedroom apartment in Rockville, MD, unpacking took no time at all. We were pretty much settled in within two days of arriving and I could refocus on in-processing Fort Bragg, a process that would encompass most of the following week.

All in all, it wasn't too bad for a two-bedroom house. After many trips to HomeGoods, she would make it into a comfortable dwelling. Thankfully, I did not have to play unit roulette at the reception company, I already knew my destiny lay with the 2nd Battalion of the 508th Parachute Infantry Regiment.

It took me only two days to in-process back into my old unit. After clearing the reception company, I immediately reported to the Brigade Headquarters of the Fury Regiment of 4th Brigade Combat Team, Michelle joining me. The 2-508th PIR S3, Major Mayo was now serving as the Brigade Executive Officer, and he gave me a warm welcome back.

There was also a new Brigade Commander after the task force had returned from Afghanistan. Colonel Tim Watson was the new Fury-6 and asked to see me prior to making my way down to the 2-Fury Headquarters. Politely meeting my future wife and introducing himself, he asked how I was doing after being away from the division. He informed me that the brigade had the Expert Infantry Badge (EIB) testing in August, only a mere three months away, and that this was a great opportunity for me to show everybody I could do anything a Paratrooper and Soldier could.

Not so eager to please, but ready to prove myself, I resolved then and there to earn that badge. After a brief conversation, I was dismissed, and Michelle and I made our way less than a mile away to the 2-508th PIR HQ.

There was also a new Battalion Commander. LTC Andrew Zieseniss had recently taken over the organization from LTC Jones. A lot of moves after the deployment it seemed, but later I realized this was completely normal in the U.S. Army. LTC Zieseniss gave us a courteous introduction and checked in on how we were managing after the transition. Explaining we were settled in and ready to work, he informed me of two significant training events over the next 90 days, a giant Joint Operations Access Exercise (JOAX) with two battalions of Paratroopers from Great Britian, and the EIB testing which COL Watson had already informed me of.

Excited at these prospects, I knew I had returned at the perfect time. Unfortunately, my wedding and honeymoon occurred directly during the JOAX. "Don't worry about it," LTC Zieseniss said, "Your wedding is the priority. There will be other events. In the meantime, focus on getting settled back in and getting through Basic Airborne Refresher (BAR)."

I never even considered it then, but looking back at it now, that was an example of great leadership. Someone who put the well-being of Soldiers and people over metrics.

Soon after in processing, I was assigned to the S3 shop to await moving down to Bravo Company, which would happen after my wedding. A week prior to the nuptials, Michelle traveled to her parent's home in Winchester, Kentucky to finalize the remaining wedding preparations. I remained behind to pass the first of many challenges I needed to overcome since donning a maroon beret again, jumping out of a perfectly good aircraft. BAR was not an issue at all for me. 1LT Pete Kavanaugh, one of the Platoon Leaders who served with me in Delta Company, completed the refresher with me prior to him being assigned to the 75th Ranger Regiment.

The day prior to the jump, my father drove into Fayetteville to witness the occasion. Having held the Master Parachutist Badge, I was sure this was a moment he could look forward to. On a Saturday in June, mere days before I drove to Kentucky to conclude my bachelorhood, I arrived at Green Ramp at Pope Army Airfield for my first jump since January 2012. To my surprise, one of my instructors and mentors from

EKU, Master Sergeant (MSG) Douglas Kleem, was jumping with me, positioned next to me on our chalk.

After watching me don my T-10 Delta parachute and undergoing the Jumpmaster Parachute Inspection (JMPI), my father traveled to the assembly area on Sicily Drop Zone to observe my exit and fall to Earth.

The aircraft was cavernous. The C-17 Globemaster can hold almost 100 Paratroopers and has ample space, far more comfortable than the C-130 Hercules. On the Jumpmaster's command "Ten minutes!" we alerted ourselves to the impending actions.

"Outboard personnel stand up!" The Paratroopers on the exterior rows of seating now stood up and faced the paratroop door in the rear, MSG Kleem and I included. At the command, "Inboard personnel stand up!" the interior seated personnel did the same and comingled with my line to form one cohesive group.

"Hook up!" Metallic clicks echoed across the aircraft as the Paratroopers hooked their static line snap hooks to the C-17's anchor line cable spanning the length of the belly of the aircraft. "Check static lines!" Using our eyes and free hands, we traced and checked our static line, the tether of life that ensured the proper deployment of silk canopies. "Check equipment!" Now we checked over ourselves and the Paratrooper to our immediate front. Everything from helmets, harnesses, and parachute components was quickly examined.

Trust in your battle buddy was imperative. Luckily, I had MSG Kleem inspecting mine. "Sound off for equipment check!" From the bow of the aircraft to the stern, Paratroopers sounded off with an "All okay!" and a smack on the backside of the soul in front of them. When you felt the slap, you sounded off and repeated the physical que, ensuring the checks went to the number one jumper and subsequently to the Jumpmaster. The last jumper bellowed, "All okay, Jumpmaster!"

The Paratroop doors opened, noise from the turbines and fresh air invading our domain. The Jumpmaster inspected the edges of the platform and door and placed most of his body well outside of the aircraft.

Satisfied with the inspection, the Jumpmaster scanned the horizon for the terrain features that marked the time until we jumped. With an index finger in the air, the Primary and Alternate Jumpmaster's signaled "One minute," which we all echoed. Then, we received the "30 seconds" cue, the thumb and index finger being positioned to signal that we were almost on target. Finally, the Jumpmasters looked at each other, nodded, and faced the Paratroopers. "Stand by!" We were there, the number one jumper now in the paratroop door waiting for the final command.

The amber light in the C-17 turned a bright green seconds later. "GO!" the Jumpmaster roared. Static lines slowly began to move forward as Paratroopers shuffled towards the door. Under a lot of weight from weapons and equipment, this can be challenging. Luckily, I was just jumping "Hollywood" or no combat equipment since this was a refresher jump.

The glow of the sun became brighter as I moved toward the door. Suddenly, the brightness became overwhelming as I handed off my static line to the safety and turned 90 degrees left. A sudden blast of noise and wind caught me as I leaped out of the aircraft. Both hands on my reserve parachute, I descended towards the earth from 800 feet above ground in freefall. After about four seconds of chaos, I felt the shock of my canopy opening abruptly, thus slowing my decent. The noise turned to silence. Looking up, I observed a circular canopy in full deployment. At least I wasn't going to hit the drop zone like a lawn dart.

I reached up and grabbed my risers, my only means of maneuvering the parachute into the wind. I saw the assembly area not more than a couple hundred meters away and slipped into the wind towards that point with a hard pull on my risers. After a minute, I was about 200 feet from the sandy surface of Sicily Drop Zone, a sprawling opening in the North Carolina pines that stretches over seven kilometers. Looking at the horizon, I saw my eyes become parallel with the trees. I tucked my chin and elbows in, deliberately ensured my prosthetic and right leg were together, and bent my knees slightly. It was all in God's hands now. If I were to get hurt, Michelle was going to kill me, I thought.

I landed with a soft thud in the sand. No pain. My leg was still attached. "Thank God!" I thought. Laying on the ground, I disconnected my canopy release assemblies and immediately began packing my chute into the accompanying aviator's kit bag. Snapping it closed, I heaved my chute over my back with my reserve to my front. My father was right there. "You alright?" he asked.

"Never been better!" I exclaimed. We walked to the assembly area together. My first test completed, I was back to being a current and qualified Paratrooper in the U.S. Army.

But now my moment was done. Michelle was next in line. The wedding was her day and now I needed to give her and the venue my fullest attention. Signing out on two weeks leave only a couple days after my successful jump, I drove back to Kentucky. After five years of engagement, this was the big day she had dreamed about probably her whole life. Family and friends from across the country were coming in for this event. I would be missing the JOAX that the division was executing, but gaining a life-long partner in my best friend. Sounds like a fair trade to me.

The wedding was held in Winchester, Kentucky on the 22nd of June 2013. Wearing my Army Service Uniform, I waited at the alter for my wife. Dozens of my friends from high school, EKU, and IBOLC were in attendance as well as many of Michelle's friends and our families. As I waited there, with my brother as my best man and my friends Brian McCrea, Adam Ward, and Caleb Wood serving as groomsmen, I contemplated the choice I made. It took little effort. After a five-year engagement, a deployment, a missing leg, and a yearlong recovery, Michelle knew exactly what she was getting herself into.

And my decision to remain on active duty despite my wounds meant that this was just the beginning. After all, I was only two years into my career. Nonetheless, she had earned the right to leave at any time. Lord knows I was not the perfect fiancé and the circumstances over the last 18 months would have tested the strongest of relationships. But she remained loyal regardless, which was all I needed.

Any doubts I may have had instantly vanished though when the hymn commenced, and I witnessed her approach the alter. She looked simply stunning. Words cannot express just how lucky I felt to know I was marrying this girl. After sharing the phrase "I do," and sealing the pact with a kiss, it was finished. I was now a married man which meant for the rest of my days, this woman and me would be bound together.

The reception was held in the Opera House in downtown Winchester. Unfortunately, due to the heat, and phantom pain in my residual limb, the venue passed by with a blur. I recall seeing my friends from college, Blake Freeman, Travis Arena, Crosby Kennedy. My friends from IBOLC, Adam Lawson and Hank Gray. For much of the reception, my new bride and I made the rounds thanking those that came, many of whom I had never met from her family's side. I never tasted my own wedding cake due to how hot I had become and fighting hard to avoid giving a hint of the pain I felt in my leg. Michelle though understood and helped me soldier on through it all, dependable as ever to help me succeed in every challenge.

Days later, we completed our honeymoon in Aruba. A tiny island located less than forty miles north of Venezuela, the resort we stayed at was a much-needed getaway for the two of us after months of being surrounded by people or couped up in the Capital Region. For over a week, we drank at a pool side bar, explored the island, and ate some fantastic food. We discussed what our future would entail. Kids were on her mind, while mine focused on an upcoming deployment to Afghanistan which we both dreaded. After much deliberation, we settled for trying to expand our family after my return home in the late summer of 2014. I did not, after all, want to leave my new wife as a widow with a newborn if that was my fate.

We returned to Fort Bragg immediately after the honeymoon which, coincidently coincided with the conclusion of the JOAX. Now the entire Fury Brigade was switching gears towards the upcoming EIB testing in early September. Two whole months were allocated prior to then for all Paratroopers to prepare for the tests. Never before had an amputee earned the EIB. I was determined to be the first and show the entire division what I was still capable of.

In early July, I was reassigned to Bravo Company as a Platoon Leader. The Company Commander, CPT Steven Robinson and I had briefly met in Afghanistan in 2012. His counterpart was 1SG Christopher Goodart whom I later found out had also served with another officer who had a missing leg, LTC David Rozelle when he was assigned to the University of Colorado Boulder Army ROTC program. Finishing out the Bravo Company Command Team was 1LT Michael Laroque, the Executive Officer who had been a Platoon Leader in the same outfit while they were deployed.

My realm focused on the four squads of Paratroopers in First Platoon. My Platoon Sergeant (PSG) was SFC Raymond Petrik who supervised the four Squad Leaders within our group. The Squad Leaders in turn led around eight to nine Paratroopers each. We had a medic, SPC McCoy; a radio telephone operator (RTO), PFC McCarrick; and a forward observer, SGT Demetrius Dasher attached to us for training. When all of the attachments were added up to the organic three rifle squads and weapons squad, the platoon had around forty individual Paratroopers coming from all walks of life.

Each platoon in Bravo Company had its own unique personality, along with their Platoon Leaders. 1LT Luke Ziller led Second Platoon while 1LT Kamal Wheeler headed Third Platoon. Other than our ranks, our backgrounds were stark contrasts. Luke, a Canadian educated officer of the Hebrew faith could be considered the most down to earth of the three Platoon Leaders with his easy demeanor. Kamal was a West Point educated officer of African American decent who was incredibly smart and fit. Serious but not too rigid to take a joke, he was easily the most well rounded of the group.

Myself, on the other hand, with my EKU education and missing leg, was probably seen as the hillbilly of the Platoon Leaders in a battalion comprised primarily of West Point graduates.

However, the differences in our backgrounds, beliefs, and appearances did not negatively impact our cohesion. In fact, Kamal and Luke became two of the most dependable officers I have ever worked with and our devotion to our duties and the organization arguably made Bravo Company one of the best units to serve in regarding culture and climate. This was just the Platoon Leaders alone. While I merely leased the platoon from the command team, the Platoon Sergeants directly managed the Paratroopers under them in terms of benefits, training, education, punishments, and taskings.

Everything that happened within First Platoon was done under the management of SFC Petrik. While I was responsible for everything the platoon did or failed to do, SFC Petrik was the leader who made things happen through the Squad Leaders. In essence, I felt more like a middleman communicating the orders and intent of the Company Commander to the Squad Leaders while SFC Petrik ensured the directives were followed through. That is the beauty of the NCO Corps. If a Platoon Leader was absent from the unit, the platoon could still function seamlessly through competent NCOs.

The Platoon Leader's role needs to set the culture of their platoon. The Lieutenant doesn't have a vision statement or leader's philosophy. That is reserved for Commanders, of which a Platoon Leader is not. He or she leads their platoon from the front regarding competence, resilience, fitness, and a myriad of other attributes and competencies.

For all of you future Platoon Leaders out there, listen up. If you are not the first one back on a run, not a problem. If you don't have a Ranger Tab, not a deal breaker. If you don't have any idea how to plan a maneuver range, someone else will know and can help. You don't have to be the smartest, strongest, or most competent officer to succeed, because I surely wasn't and did just fine.

Success isn't based on your achievements and merits, but on the performance and readiness of the organization you lead. In essence, a devoted leader who cares for the well-being of their Soldiers with no Ranger Tab is hands down better than the arrogant officer who lives off their own laurels. Soldiers at every level see through the B.S. Show them you care by following through and holding up your pact to serve them. By focusing on your platoon's performance and readiness and actively showing an interest in the health and welfare of the Soldiers under your charge, the men and women serving under you will make you successful.

Ending a five-year engagement (JUN 2013)

CHAPTER 13
LEARNING TO FIGHT AGAIN
AUGUST TO OCTOBER 2013
FORT BRAGG, NORTH CAROLINA

The focus of the entire company at this point was getting ready for the EIB. In the former 4th Brigade Combat Team footprint, there is a wooded area that separated the housing area from the unit facilities. Known as the "fishbowl" to those that were there, this was where the bulk of the EIB training was conducted. Inside this ad hoc training area were dozens of little stations, each comprising some type of weapons system or piece of equipment that trained a task that was to be tested. A Paratrooper who had earned the EIB was assigned to each station and helped instruct their audience on the performance measures of the task to be trained.

The entire training area in the fishbowl was resourced perfectly. It was adjacent to the unit barracks and headquarters which made it very accessible. All the equipment and weapons were resourced internally, and the cadre were all competent NCOs who standardized the training. The tasks were the heart of the EIB testing and varied from ridiculously simple, to intricate and difficult to master without repetition. But that was what this was all about, repetition. Paratroopers had hours every day to practice the performance measures until they were second nature. This was one of those instances where you get out what you put in.

Apart from the tasks, every Paratrooper competing in EIB qualification needed to have an expert qualification score of 36 out of 40 on their assigned weapons system, the M4A1 Carbine. Thus, First Platoon, along with the rest of Bravo Company was at one of the dozens of ranges on Fort Bragg sending steel downrange. My PSG, SFC Petrik was one of the best shots in the platoon and his guidance helped ensure the Paratroopers' success. By the end of one week and a lot of bullets expended, every member had achieved an expert score, and thus had their ticket to September's EIB showdown.

Throughout July and August, I also endeavored on a side project of getting caught up on my jumps. After being off airborne status for over a year, I only had participated in ten airborne operations. To make it to Jumpmaster School, a Paratrooper needs 12 exits from a C-130 or C17 High Performance Aircraft. I was determined to get caught up.

Talking with the Battalion S3 Air Team, 1LT Vanderlip and SFC Jenner, I obtained the jump schedules for the battalion and the rest of the division. In my free time, I volunteered to accompany other jumps with the 82nd to pad my jump log. Known as "strap hanging" I literally showed up to the Initial Manifest Call (IFC) of other units and informed a Jumpmaster that I was trying to jump. Most of the time, chalks had a vacancy so I could simply add my name to the roster and participate.

Within two months, I added five more jumps to my jump log, to include my first night jump with combat equipment. Giving up one Saturday, I managed to secure my first foreign wings jump with the Civil Affairs unit on Fort Bragg. Hosted by Chilean Paratroopers, the jump was out of a CH-47 Chinook helicopter on St. Mere Eglise Drop Zone. While rotary wing jumps are few and far between, the real rarity was the foreign wings which I could now proudly display on my Army Service Uniform.

However, this may not have been the wisest distraction from my duties as a Platoon Leader. A bad landing could easily result in a broken leg or concussion. As an amputee, I could ill afford the loss of another limb, let alone become injured on airborne operations that were not necessary for me to execute to remain current. Luckily for me though, I made it through these extracurricular jumps without incident and always returned home to my wife unscathed.

The week prior to EIB testing, the entire brigade trained at one of the training areas for a final round of preparation. At Area H, additional land navigation was conducted along with training lanes set up to mimic the testing conditions. Once more, the Fury Brigade really set the example for how to do a proper train-up.

Unfortunately, a Paratrooper conducting land navigation training managed to misplace his PEQ-15 laser. Somehow, the laser detached from his M4A1 Carbine. For two days, all Paratroopers on Area H lined up abreast and walked within arm's length of each other to find the laser. These two days could have been spent training on the tasks, the meat, and potatoes of earning the EIB. Thankfully, one of the Paratroopers from Bravo Company found the laser and thus, ended the "Hands Across America" spectacle.

Still feeling like I needed additional training on the weapons tasks, I coordinated with the unit armorer to come into the B Company Operating Facility (COF) on a Saturday to train on the weapons tested during EIB. I broadcasted this opportunity to the other Paratroopers, but only a handful showed up.

A MK-19, M2 .50 Cal, M240-B, and M249-SAW were brought out of the vault and assembled. For hours, a half dozen of us went through the performance measures of the tested weapon until they were second nature. Our sequence improved tremendously, and our times went down considerably. By the end of the afternoon, I felt more than confident. I had given it everything I had to ensure I was ready for this challenge. I was in outstanding shape. I was very comfortable with my land navigation skills. I was able to ruck without issues due to the miles logged with my platoon and during my free time. There would be no other time in my career when I was this ready.

On that first Monday in September, EIB testing kicked off on the 82nd Airborne Division's parade ground, formally known as Pike Field. The breakdown of EIB was as follows: Day one was the APFT on Pike Field and two bouts of land navigation at the Area J training area (day and night respectively).

For those who passed these events, the next two days were the actual testing lanes. Candidates had three lanes to pass, each one containing different tasks while mimicking a tactical patrol. Prior to the initiation of a lane, the candidate had three lane initiation tasks that they needed to successfully perform. If a testee failed any one of the tasks twice, they were eliminated from EIB. You could execute one or two of the lanes in one day and complete the remainder on the second. The fourth and last day was the twelve-mile ruck march and, upon a successful completion, graduation.

I passed the first event without any concerns. The APFT, two minutes of pushups, sit-ups, and a two-mile run with 70 percent minimum scores, should have weeded out only a few Paratroopers. Instead, I remember a crowd that must have been over a hundred failures on that vast field. I distinctly remember the Brigade and Battalion Command Sergeants Majors nearly losing their minds at the gathering of eliminated candidates.

Choruses of "How in the hell could you fail an APFT?" and "You all are a damn disgrace!" echoed across the field. One of them even pointed at me and said, "You see that! A one-legged man just beat you!" Not wanting to draw any attention to myself for just existing at that moment, I retired from the parade field to shower and change into my Army Combat Uniform (ACU) for land navigation.

Back at the COF, I regained my focus for the land navigation event that was to start in a couple hours. Even though it was going to rain later, I forced myself to shower prior to going out into the field just to get grimy. MSG Douglas Kleem's advice from ROTC was in my head at this time. "Whenever you have a chance to shower, take it." That wisdom paid dividends over the last couple years from IBOLC to Ranger School. No matter how tired you are, get clean. Even though I wasn't going to catch any rack anytime soon, a hot shower helped me prepare my mind for the evolutions ahead.

Land navigation that day was at Area J, the primary course for most Paratroopers in the division. Located adjacent to Longstreet Road and just outside the installation cantonment area, the course is mainly wooded with easing topography changes throughout it. Running through it like a nervous system is a network of trails and fire breaks that show prominently on the Fort Bragg Special map.

Arriving about an hour early to the assembly area, I took a moment to survey my surroundings. Tents and camouflage netting served as the site's operations center for the land navigation committee. Sitting on a collapsible camping chair with my M4A1 carbine next to me, I observed candidates starting to march in. Some of them were members of Bravo Company and my platoon, but most were new faces. As the minutes winded down until the commencement of the training, the air became noticeably cooler, a tell-tale sign that rain was imminent.

The instructions were simple. Using a map, compass, protractor, and in the assigned uniform, find four out of five points in two and half hours. No phones, ground positioning systems, or smart watches were authorized. Usually, a GO in navigation at Area J was three out of five points in three hours. But not if you wanted the EIB. I had to move quickly and efficiently, I realized, if I was to make it. Luckily for all candidates, roads, trails, and fire breaks were authorized and fair game. Instead of a regular prosthetic foot, I utilized a running blade to allow me to move quicker. This technique would later become standard for me in the duty uniform as an Infantry Officer.

After the safety brief and final instructions, a new riddle for the candidates to solve was thrown at us. All candidates were given an eight-digit grid for their start point and had to not only find out where that point was to begin, but had to further convert the eight-digit code into ten digits to further refine the accuracy of the position on the ground. Though I forget the details now, the ten-digit grid had to be within a few meters (between five and ten) to be considered accurate enough for EIB. This riddle was a pass or fail event needed to move on in EIB, in addition to the four other points one must locate. On the command "Begin," the clock started running. Rain had now begun to fall on that muggy day. In addition, to my land navigation accessories, I had my weapon and kit to accompany me.

I wasted no time plotting my five points on the map. I double checked my work. For this task, I utilized my own protractor. My good friend, Major Franzen, a Green Beret, and a hell of a human being, gave me some wisdom prior to land navigation. He advised me to get a protractor that had the 1:50,000 and 1:25:000 measurement stencils finely grafted. In this way, when plotting, a quarter centimeter of plastic did not obstruct you when marking your points. Even a distance that small could wreak havoc during land navigation. Especially if two points were located near one another and only one was the right location.

I looked at the protractor in the kit bag. It was one of the generic store-bought protractors with just such a handicap. Taking out my own, I plotted each point twice, more confident in my own skills now. As long as I turned in their protractor at the conclusion of the event, there was no harm or foul.

Now I had to determine where the start point was. Every 50 meters from the intersection of Gruber and Longstreet to the entrance of the land navigation assembly area was a metal point on a pole. Three-sided orange and white coded signs, these points were the easiest to find. However, under or overshooting your pace count could lead a trainee to the wrong position and thus, heartbreak. After plotting my start point, I plotted the distance from the Area J entrance off Longstreet to it.

In this capacity, I knew just how many paces to take in order to not overshoot the point. I was confident in my pace count. I had practiced it at Area H the previous week. 64 paces off the right foot marked every 100 meters. My ranger beads served as my abacus to ensure I kept up my math on longer distances. A bead moved up the 550 cord every 100 meters. After 1000 meters or a kilometer, I moved a bead on a separate line and moved the other nine beads back to the bottom to restart. Some think the ranger beads are for ROTC nerds, but they have never let me down.

Almost twenty minutes had elapsed by the time I reached my starting point. The rain was falling steadily now. I needed to move quicker. I took my map out and replotted the point. Converting it from an eight to ten-digit grid was simple enough for me. I double checked my work and moved on, confident that I was on the money. Taking out my waterproof score sheet. I wrote the answer to the equation on it and prepared for the movement to my first "real" point. I had under a kilometer to move and decided to utilize the intersections of the fire breaks as attack points. By doing this, all I needed was the map. The compass, secured to my kit, was to be used to shoot an azimuth once I was within 50 meters of my point.

I marched brisky through the woods to my first location. There were no trails parallel to my route so breaking brush in a straight line was the best course of action. There was one terrain feature that could possibly hinder me, a creek and thus, a draw that was no doubt heavily vegetated with undergrowth. I have heard horror stories from the Special Forces Assessment and Selection candidates at neighboring Camp Mackall regarding the draws of the infamous star course. As luck would have it though, I trekked through it without much delay.

Coming across a major firebreak on the map, I turned right and moved a couple hundred meters to a large intersection. I took out my map. I was about 50 meters away from my first point. Shooting an azimuth with my compass, I walked into the woods. A small beaten trail signaled that I was on the right track, the path utilized by hundreds of trainees prior to me throughout that summer. Within a few minutes, the path stopped at a three-sided point on a pole. I took out my score sheet and wrote down the code. To further prove that a candidate made it to the point, a clacker was hung from the pole by some 550 cord. I clacked the score sheet, little holes now showing another code from the pins in the clacker.

I had less than two hours to go to obtain three more sets of codes to be considered a GO. The rain, however, was beginning to work against me. The prosthetic blade, while awesome on pavement, was sliding on the sandy firebreaks which now were becoming slick. By using this leg to move quicker on the course, I was now becoming hindered by the mud and sand that sloshed under me.

I plotted my next distance to the second point and moved out, sliding down the trails as the incessant rain fell even harder. I found my second and third point without much hardship, but the traveling time took its toll. I now had less than 30 minutes to find the fourth point and return to the assembly area to turn in my score card. I was well over one kilometer away from my fourth point and over two kilometers from the land navigation committee. Thankfully, my fourth point was manageable, being routed back to the start.

I double timed now. If I walked at all, I was done for. My weapon at times helped me keep from busting my fourth point of contact on that godforsaken trail that may as well have been greased over. The rain was now a torrential downpour. Sliding, falling, and forcing myself forward, I closed the distance. In 15 minutes, I reached my fourth point in a clearing. I marked my score sheet in the same manner as the other points and quickly checked my bearings. I had no time to lose now.

With only ten minutes to spare, I had to make it back to the start point and turn in my score card. I did one last map check. I could sprint straight down the trail to Longstreet and then move to the start point. It wasn't the shortest distance I could have taken, but the hardpacked vehicle trail would enable me to move quicker as opposed to the muddy and sandy fire breaks. Checking my kit to ensure nothing was out of the ordinary and holding my carbine, I took off.

With less than ten minutes, I moved as fast as I could go. My left limb hurt like hell from the sliding, and I could tell I was going to be in some pain once the adrenaline subsided. It didn't matter at that moment though. The rain fell harder on me. My right boot was drenched to the foot and the hot spots from the blisters were heating up. I saw the exit from Area J to Longstreet ahead and pushed forward. I had five minutes to go. Turning right, I sprinted down the trail parallel to the hardball road and closed the final distance. The mud on my boot and the saturated clothing and kit weighed me down.

Looking at my Casio G-Shock on my left wrist, I saw I had just two minutes left as I turned into the land navigation assembly area. Candidates milled about. Some tried to stay dry under ponchos while others simply embraced the elements. With a minute to spare, I arrived at the tent. I handed in my scorecard to a member of the lane committee for inspection and grading.

My heart palpitated as the NCO checked my ten-digit grid and the other five points on my sheet. A horn sounded the end of the day land navigation evolution as he judged me. I literally had no time to spare I thought. With great relief, I noticed that the red side of his pencil stayed upright, only the blue side marked my paper to show I had found the right coded points. "Four out of five. You're a GO," he informed me. "With no time to spare," he added. That was that. My second test completed; I could breathe a sigh of relief.

The committee leader soon beckoned all the surviving candidates forward for further instructions. Our group was severely attritted by the APFT and day land navigation alone. No doubt more victims would accumulate during the next field problem, night land navigation. We were instructed to arrive back at the training area by 2000 hours. It was midafternoon at about this time, so I decided to return to the COF, secure my weapon, and repeat the process of showering and getting into dry clothes. The COF was over a mile away from the location and vehicles were not permitted on the site. In the rain, sore and in pain from blisters on real and imagined feet, I trudged back to my home at Bravo.

Day one of EIB had cut the number of candidates in half and we hadn't even completed the day's final event. Most of my platoon was out of the fight at this point, along with much of Bravo. I resolved to not lose this battle. I applied ample amounts of foot power to my remaining foot and Vaseline on my left limb to reduce the friction with the prosthetic sleeve from sliding. That afternoon, the weather had cleared up, reducing the likelihood of becoming drenched in the period of darkness when all focus was on navigating the pitch-black course.

Wandering through the same woods as before, I quickly found my first point. The same rules applied as previous for day land navigation except now I had to find three out of five points in under two hours. There was no initiation test like the grid conversion, so we were free to plot and move out from our starting positions. I was authorized the use of a flashlight with a red lens on it during stationary pauses but not during movements. Instead of carrying a flashlight, most in the Army utilizes a head lamp. My Petzl headlamp, the same one that accompanied me through IBOLC and Ranger School, was still going strong with a fresh trio of AAA batteries. I was careful to avoid leaving the trails and fire breaks; breaking brush is hard enough during the day but it's a whole different story during limited visibility.

I gambled during my pause between day and night land navigation to retain my blade in the hopes the trails would dry out. Although still slick, I was able to run without too much sliding. I quickly found my last two points in quick succession. The beautiful thing about attack points was that if you know how to terrain associate very well, you hardly need a compass. Know where you are, plot checkpoints on your map to ensure you keep your bearings, but most of all, be confident in yourself.

The first day of EIB thus ended for me with some great training and three victories. I was still in the running for the badge. I was dog tired from the amount of movement that day. Between the APFT and two bouts of land navigation, I needed to get some rest. Perhaps most important though was getting off my prosthetic and giving my limb a rest. The next two days were more mental than physical for the remaining candidates. We had three lanes of tasks to execute flawlessly. Typically, these two days weed out a considerable number of candidates. Still though, I felt confident after practicing the tasks until I couldn't make a mistake.

The next day, the remaining candidates were trucked by LMTV to Area H, the same location as the practice training we received the week prior. There were three pits that the candidates could congregate at. One pit had all the weapons that would be tested for the candidates to practice tasks' performance measures. The second pit had medical supplies and Rescue Randy's (mannequins) for practicing medical tasks. The last pit had communications, CBRNE, and miscellaneous task equipment. However, there were no NCOs to guide you this time. Either you knew it, or you didn't.

The EIB committee warmly welcomed us and congratulated us for getting this far. We were told that we had two days to execute the lanes. The first lane for everyone was chosen for them to ensure every candidate went through one lane and bottlenecks didn't occur. I was selected for the Alpha Lane for my first round. Just as with day land navigation, there was a series of initiation tasks that needed to be completed prior to entering the lane. The three tasks were placing a AN/PRC-148 MBI-TR into operation and uploading encrypted keys utilizing a Simple Key Loader (SKL); loading, firing, clearing a jam, and putting an M4A1 back into operation; and taking apart and reassembling the M9 Berretta in under thirty seconds. You were allowed to fail one task. If you failed two on one lane, you were eliminated. Throughout the entirety of the EIB tasks, failing a cumulative two tasks resulted in dismissal from the testing.

Right off the back, I made a mistake when I initiated my turn. I missed a step in uploading the encrypted keys into the MBITR from the SKL. The SKL that day was anything but simple to me. I agonized as the grader's red side of the pencil marked my score card. If I failed the M4 or M9, I was done for. Quickly ignoring the setback, I pounced on the M4. After firing two blank rounds, it jammed due to an inert blank round. MSG Dick Sirry, a former ROTC instructor from EKU had taught me how to remedy this long ago. SPORTS is the acronym for clearing a jam. Slap the magazine, pull the charging handle to the rear, observe the chamber, release the charging handle, tap the forward assist, and squeeze the trigger.

I executed this to the standard and breathed a quick sigh of relief as the weapon fired two blank rounds. Clearing the weapon and placing it back on safe, I laid the M4 on the table and announced "Done!" to the grader. A blue mark hit the card and I was now faced with the easiest task, in my opinion. Within 30 seconds I quickly cleared, disassembled, assembled, and conducted a functions check on the M9. I had practiced dozens of times previously and was extremely confident in my execution. Once again, the blue pencil marked home.

The details of the tasks of the Alpha Lane escape memory, but whatever was the case, I ended up passing and being cleared for the next lane. I decided to execute the Bravo Lane that afternoon after taking a break. I had all day to execute two lanes and only those eliminated were allowed to leave the training site for the time being. Going to one of the pits, I took off my leg and rested on my collapsable chair. I studied some of the other tasks such as call for fire and range estimation while taking moments to meditate and clear my mind.

After an MRE for lunch, I moved to the site of the Bravo Lane. The three tasks of the Bravo Lane consisted of putting an AN/PVS-14-Night Vision Device (NVD) into operation; loading, firing, and clearing the M249-SAW Light Machine Gun; and executing the corresponding functions checks to the MK-19 Heavy Grenade Launcher.

You had 10 seconds to place the NVD into operation. It was simple.

Take the NVD out of the bag, put the batteries in, and turn the switch on. That's it. After executing this task, I moved on to the M249-SAW. This was one of the more difficult tasks. Within 30 seconds, a candidate had to clear the weapon, load the weapon, fire the weapon, and clear it again. I was given about five blank linked rounds of 5.56mm for this task. The trick though is to tilt the SAW so that the rounds remain on the feed tray when you close the feed tray cover. Otherwise, they would simply fall off or the weapon wouldn't function correctly if they jammed. I practiced this in the B CO COF for what seemed like hours before I was content with my sequence and timing. The practice that day paid off when I knocked it out of the park.

However, the final task prior to the Bravo Lane about sent me home. I goofed up on the performance measures of the MK-19 by emplacing the weapon on safe one too many times. The sad thing is, I knew I messed up while I did it. The grader gave me the option to either try again after remedial training or to execute the task again on the spot. I told him I wanted to try again right then and there, much to his chagrin.

Luckily though, I nailed it. Now I was blade running. I couldn't miss any more tasks throughout the remainder of the lanes. The Bravo Lane consisted of several tasks. Three of them were conducting individual movement techniques, employing hand grenades, and calling in a 9-Line MEDEVAC Report. It wasn't difficult by any means with the exception that my prosthetic almost came off after bounding through the lane. Within the allotted ten minutes, I completed the lane to the standard.

Turning my scorecard in to the grader, I was authorized to leave the training area since I had performed two lanes that day. I could stay and practice or get some rest. I chose the latter being that the hardest tasks for me had been accomplished. Tomorrow was the patrol lane and following that was the 12-mile forced march. Walking back to the COF across the cantonment zone, I announced that I was still in the running to the members of the platoon who were out of the running and still on duty.

A chorus of cheers went up from those Paratroopers, which lifted my heart. I informed CPT Robinson and 1SG Goodart of my status, to which both seemed pleased, and turned my weapon into the arms room. Driving back to my rental at Piping Plover, Michelle was eager to know how I was doing. Ecstatic that her husband was proving he was just as good as some of the best Paratroopers despite one leg, we settled in after I took a shower and mentally prepared for the next day's challenges.

The final lane was executed without mishap. The pre-execution tasks consisted of putting the M2 .50 caliber Heavy Machine Gun and M240-B Medium Machine Gun into operation followed by a distance estimation test along an enormous fire break at Area H. The distance estimation test was aided by the fact that power lines ran parallel to the break in the wood line; providing a quick reference to the distance need- ed to be estimated within a small margin of error.

The Patrol Lane was designed to mimic combat conditions, albeit in lieu of a squad, you were alone. I started the lane by moving along a path until encountering a casualty. The victim, a Rescue Randy, had wounds that had to be tended to in accordance with Combat Lifesaver Skills (CLS) training. You do not need to be a surgeon to figure it out. First, assess the casualty. Check for alertness, whether he/she is uncon- scious. Then, because a dummy can't respond and is thus, unconscious, check the airway and pulse. Next, find the wound. Is there blood? Yes? Apply a tourniquet to an extremity and a pressure bandage. Call up the 9-Line MEDEVAC. Done. Well, there is a bit more to it, but you get the picture.

The remainder of the lane was oriented around a Chemical, Bio- logical, Nuclear Environment (CBNE) attack. Discovering indicators of some type of chemical gas attack, I quickly donned my pro-mask. Check. A device which detects chemicals was situated on the lane. The last obstacle left of the final lane was putting it into operation. I ensured the testing strips were loaded, turned a couple dials, and boom. Good to go. The observing grader called ENDEX. Removing my mask, the NCO told me I was good to go and to report to the EIB Committee Operations Center for further instructions. It was that simple, yet the buildup made it seem like this lane would be the most difficult of them all. I guess the training paid off.

I was probably floating on air. All I needed now was to pass the 12-mile forced march. Rucking was a skill I excelled at since college. I had completed multiple marathon ruck marches in the Smokey Mountains during my time in ROTC. Even after losing my leg, I could still break 15-minute miles easily with a little shuffling. I was congratulated by an officer in the tent for my performance thus far. Another NCO advised me to report the next day at 0200 hours at the 2-Fury HHC COF. My weapon would be secured for the rest of the day and overnight so that the armorer didn't have to lose sleep for me.

That afternoon, I ate an entire pizza from Pizza Hut. I needed the carbs for the ruck and the sleep. I was confident and not too concerned. Over that summer, I had been introduced to GORUCK by MAJ Marcus Franzen. GORUCK in the plainest terms tests people's character, physical fitness, and patriotism. It is not a Tough Mudder or Spartan Race but an hours long tribulation where teamwork is essential to the accomplishment of the event.

Starting off with a "welcome party" where participants are "smoked" for a couple hours with some exercises, the event then moves into a long movement that typically involves carrying some random heavy objects such as logs, water cans, or tough boxes of weights. I have seen instances where boats are even carried. Participants will have a 30-pound bag or sack on their back for the entirety of the event. For completion, participants are awarded a small patch with the GORUCK logo on it. 1LT Luke Ziller, a fellow Platoon Leader with me in Bravo, even accommodated me on some of the events. After conducting three GORUCK's over the summer, I felt confident and ready to smash these 12 miles.

The route was six miles down Longstreet and six back. However, the path itself wasn't a street by any means. Past the cantonment area entry control point, the paved road turns into a two-lane hard packed gravel road. The further west you marched, the rougher the route became as gravel loosened up. The start point was pitch black except for the LMTV that chauffeured our small band. After four days of competing for the EIB, the 4th BCT of the 82nd had slighter more than a dozen candidates left. The lanes had taken their toll as expected. The route itself had no lighting. Then again, it was a long six-mile stretch; no land navigation was required. The turn around point was marked by some chem lights and a HMMWV with two cadre recording names.

After the command "Go" was announced by the 2-Fury Operations Sergeant Major, SFC Jack White, the candidates began their trek into the pitch-black void. I decided on a tried-and-true technique for achieving the 15-minute time hack necessary to complete the 12-mile forced march in under three hours. I would jog for one minute and walk the next minute. For six miles, with 35 pounds on my shoulders and back, my helmet donned, and my weapon in my hands, I did just that. The weather was warm but had begun cooling off due to the commencement of the fall season. I had no companions but my thoughts and my pace.

At the turn around marker, I was joined by the 2-Fury Chaplain. A good-natured Captain whose name I regretfully fail to remember, he politely asked if he could march with me. "Absolutely," I replied, thinking I needed as much of God's grace that I could get from the soreness of the left limb setting in. Luckily for me, I came in to the six-mile marker under an hour and fifteen minutes. I could theoretically walk the remaining six miles and still come in under three hours. As we closed the distance, we shared our stories about our families and adventures. The whole time I considered that this was truly a great leader and chaplain. I mean, he only woke up at 0200 hours just to walk with the candidates and provide them companionship and spiritual guidance. To my recollection, no other leaders were out that day.

Finally, though, after almost three hours of walking on the hard packed gravel of Longstreet, I saw that the end was near. SFC White was standing at the finish line, annotating on his clip board those candidates that had already passed. Some were ahead of me, others to my six-o'clock. When I was within 20 meters of finishing, I jokingly yelled, "I'm ETS-ing!" An acronym meaning End of Terminal Service or departing from the Army.

SFC White quickly retorted with, 'Well ETS your ass across the finish line." And so, I did. With less than five minutes to spare. It was over. I had completed the necessary tasks needed to obtain the coveted EIB. While happy, I was sore and feeling broken down. I needed a break. Dropping my ruck, I loaded the LMTV and about passed out when I removed the prosthetic. The smell was horrendous from the socket due to the sweat. But it didn't matter. I went the distance and came out victorious, on one leg.

The rest of the 82nd Airborne Division was also busy that morning. The XVIII Airborne Corps had a run with all participating units to include the entirety of the 82nd. While they were still jogging down Ardennes and Longstreet, I collapsed in my office. I tried to fall asleep and await the arrival of my platoon from PT, but to no avail. The adrenaline was still pumping throughout my body. I called Michelle and told her the news of my triumph. She did not have a doubt in me as usual.

Suddenly, the COF erupted in noise as the Paratroopers of Bravo Company entered the building after a fun four-mile jog with almost a ten thousand other Paratroopers and Soldiers. 1SG Goodart entered the First Platoon office where I dozed and seemed to shake me back to reality from my semi-conscious state. "Did you pass?" he inquired. Shaking my head north to south, he erupted with a "Fuck yes!" and exited the office as quickly as he entered, obviously pleased one of his own had passed the exams of EIB.

The entirety of the 4th Brigade Combat Team stood in formation that afternoon. Thousands of Paratroopers stood in their summer blues uniform for it was Pay Day Activities today. All except eight Paratroopers who wore their combat uniform. Eight Expert Infantry Badges were awarded that day, two for Bravo Company Paratroopers of the 2-508th PIR. As COL Watson pinned the metal badge of the silver long rifle on an infantry blue background to my left breast, I felt a sense of relief. "Proud of you," the Brigade Commander said approvingly. I had passed my first real test coming back to the line. I was dog tired, but thankful it was over.

Being the first amputee to earn the EIB, I now contended with a new dynamic that prior to this day, I was unaccustomed to… public relations and media. Immediately after the ceremony, the Division Public Affairs Officer asked if she could conduct an interview, to which I agreed. Soon, I was sitting in podcasts, on the local radio, and was on social media and the internet. I wasn't accustomed to the attention, but it almost felt natural just to be myself.

I didn't realize it at the time, but I was given a platform for my voice to be heard. It wasn't just my voice though. Every wounded Service Member, civilians with physical disabilities, and Soldiers struggling with their own resiliency needed a champion. Talking later to some of my friends from Walter Reed, they encouraged me to tell my story and get it out for awareness. This was truly when I understood that this wasn't about me anymore. If I was to represent an entire community, I had to be flawless and strive to perfect my character and competence. From this time on, everything I did as a Soldier, I ensured that the legacy I was continuing for those that could no longer serve was carried with honor.

Day One of Expert Infantry Badge Testing (SEP 2013)

CHAPTER 14

RETURNING TO AFGHANISTAN
OCTOBER TO DECEMBER 2013
FORT BRAGG, NORTH CAROLINA

From the remainder of September to November, the 2-508th PIR showed no signs of slowing down due to their imminent deployment to Afghanistan in support of Operation Enduring Freedom XIV. The rest of September geared towards a three-week field training exercise (FTX) on post since the Joint Regional Training Center rotation was cancelled due to the government failing to pass the budget; something no Paratrooper was complaining about. Thus, for the remainder of September and most of October, Bravo Company trained for future missions in Afghanistan. Battalion airborne operations, company air assaults, and patrolling lanes at the platoon level helped First Platoon refine its TTPs and SOPs.

Due to the force capacity of the theater, the Battalion Task Force was small in comparison with the 2-Fury deployment of 2012. Instead of all units deploying, some were left on rear detachment to include the Forward Sustainment Company. Bravo Company's mission allowed for three platoons but with only two squads each to accomplish our mission, which we would find out was entirely different than what was performed over a year earlier.

On the home front, transitions were also happening. Michelle and I purchased our first home in late September; literally signing papers and obtaining keys the day prior to the battalion commencing it's three-week FTX. This later proved to be one of the best financial decisions I could have made during this era of my career and life. A beautiful brick two story home with plenty of space for a couple and a puppy looking to expand. The neighborhood was also a lot nicer and more established than the sprawling subdivision we just vacated. The office was only fifteen minutes away with Honeycut Gate only five miles away. Located in Pine Forest, it was just north of Fayetteville and east of Spring Lake. And the best aspect was that I was no longer paying off someone else's mortgage, but building my own equity.

Due to the FTX, Michelle was responsible for moving our possessions from the rental property to our new home. Luckily, she wasn't left to figure it out. Two miles away lived SSG Retired Eric Myers, a fellow Fury Paratrooper who had lost both of his legs from the waist down on the 2012 deployment to Kandahar. With his wife Laura and young daughter Kinley, we became fast friends at Walter Reed and attended many of the recreational therapy trips together.

Answering the call for support, the Myers family recruited their neighbors and brought their trucks and trailers over to assist Michelle. Within a day, all of our property was relocated to 7630 Trappers Road and Piping Plover was vacated. The military community is tight knit. Spouses and Service Members alike typically support each other through thick and thin. Fayetteville is no exception to this rule.

While Michelle was turning a house into a home, I trained with the Paratroopers in First Platoon. CPT Robinson had Bravo Company air assaulting into the training area to kick off the FTX. Throughout the evening, CH-47 Chinooks inserted platoons of Paratroopers into a landing zone. With over eighty pounds of weight and a spare prosthetic,

I moved off the bird; nearly blowing my knees out as I stepped off the tail gate of the school bus sized helicopter. Recovering, the First Squad Leader, SSG Duncan Kiruthi, led the four squads of the platoon to our patrol base location about two kilometers away. With SFC Petrik bringing up the rear of the movement to ensure there were no stragglers, we settled into the woods adjacent to a clearing that served as the Company Command Post. The evening was perfect and within no time the platoon was ready to settle in after conducting their security plan and priorities of work.

For three work weeks with weekends off to save on funds, the Paratroopers of 2Fury executed training from airborne operations, air assaults, counter improvised explosive device (C-IED), platoon and squad patrols, and received training from contractors on new equipment fielded in theater.

Overall, the weather was perfect for a North Carolina fall and there wasn't too much to complain about. The culminating event was the platoon live fires at Observation Post (OP) 13. The mission was simple. With a platoon sized element, assault and clear a compound utilizing organic fires. For a whole day, the platoon rehearsed for the mission. We built a terrain model of the entire objective utilizing old dunnage boxes and MRE trash. With the three Rifle Squad Leaders and the Weapons Squad Leader, SFC Petrik and I went over the scheme of maneuver until every leader knew what to do. Special teams rehearsed enemy prisoner of war or aid and litter drills. We were confident in our abilities, and I felt the platoon was ready to tackle the mission.

The Battalion Commander, LTC Zieseniss was walking every platoon lane with the Battalion Command Sergeant Major, CSM Kelly and the Company Command Team.

When we initiated the assault, the weapons squad at the support by fire position was making the guns sing. Soon the rifle squad assigned to the breach created an opening into the compound. I followed the assault in as they systematically cleared the structures within the compound. I observed as my guys cleared each room and eliminated any enemy targets. But I forgot one very important consideration as a Platoon Leader.

LTC Zieseniss and CPT Robinson were quick to let me know that by being in the structures on the compound with the lead trace element, I had essentially focused on one dynamic of the mission as opposed to the big picture. Had I moved to a position within the compound that allowed me to observe its entirety as opposed to following in the assault squads, I could see what was happening everywhere and paint a better picture for the Commander. Humbled, I ensured that I backed off the clearing teams during the next iterations and positioned myself where I could better lead my platoon.

After nine rifle platoons had validated at OP-13, one final evolution remained for the Paratroopers of 2-Fury. A long ruck march through the training areas to the All-American Landing Zone (LZ) for extraction by the same CH-47's that brought us in.

Throughout that October night, 1LT Luke Ziller's Second Platoon took lead and navigated the company through sandy fire breaks and across giant drops zones. The march had to have been less than ten kilometers, but it felt like an eternity due to the slow pace and the short halts to check navigation and allow stragglers to catch up.

To this day, I despise ruck marches where I can't control the pace. I prefer to move fast, and a slow pace aggravates the limb rubbing in a prosthetic. In short, rucking hurts so I move quickly to get it done as quickly as possible.

Towards dawn, the company arrived at the LZ where we moved into chalks for pickup zone (PZ) posture. The plan was for the birds to pick us up and fly us to a closer LZ where we would move back to the battalion footprint as a cohesive unit.

As fortune would have it, the landing zone was adjacent to Area J so the route would be considerably shorter. After a short flight of roughly ten minutes, the giant helicopters put us down and the Paratroopers of

CPT Robinson's Bravo Company hastened to form the order of movement so that we could leave the FTX behind us.

After another three miles of walking, we entered the 4th Brigade Combat Team area. The Boys are Back in Town by Thin Lizzy was blaring from loudspeakers. The Family Readiness Group (FRG) was preparing and serving breakfast to the returning Paratroopers. It was a welcome sight for the team and served as a well-deserved departing gift for completing our last training exercise prior to the deployment.

For the remainder of October and well into November, Michelle and I made the best of our short time together in our home prior to my departure. She had already put her touch on the interiors of our residence, quickly turning it into a warm home that echoed her personality. I thought I was lucky. Michelle had outstanding taste in interior design. There was no clutter. No unnecessary decorations or gaudy displays anywhere. She focused on the rustic aspects of design and brought a little piece of home from Kentucky in her efforts.

Most importantly though, she left a room across from the master bedroom upstairs empty. That room was destined for when I returned from my second tour and the family's planned expansion.

Not to be outdone, I added my own touches to our home in the Pine Valley area of Fayetteville. I leveled a portion of the back yard; first by hand and then by Bob Cat machine. Then I had a concrete patio and fire pit poured to make a larger outdoor entertainment area. I walled off a retaining wall with some decorative rock before finally building my first home project, a pergola over the entire patio. To finish off the backyard, I built in two horseshoe pits so that my father and I could play.

For a 24-year-old, I was doing pretty good for myself, I thought. My first investment in real estate would hopefully pay off. Now I had to live in the house long enough to make any money off it.

My longtime friend, Scott Stafford, also began his career at Fort Bragg. He was commissioned as a Field Artillery Officer and assigned to the 2nd Brigade Combat Team not long after Michelle and I arrived back at Bragg from Walter Reed. As luck would have it, a house on my street was up for rent that would support Scotty and his two large Alaskan Huskies. His presence was awesome that fall as having another friend of Michelle and me around forced us to entertain more.

Prior to the upcoming deployment, I hosted a cookout for the entire platoon and invited the rest of Bravo Company. It was a full turnout with the Company Commander even showing up. One of our previous Squad Leaders, SSG Turner, even brought over a half dozen jars of moonshine, of which contributed greatly to the party's lively atmosphere of grilled meats and jolly Paratroopers.

Most of my time as a Platoon Leader for the remainder of the month consisted of finalizing the platoon's equipment load out. Small containers, called Quadcons, were positioned behind the unit COF for loading equipment into prior to being transported to the coast for seabound transport to the Central Command (CENTCOM) Theater of Operations. First Platoon was taking a section sized element of its best Paratroopers divided into two squads led by SSG Daniel Dreeson and SGT Jacob Tippit.

Throughout November, the Company Command Team and the platoon leadership received briefs on their Area of Operations (AO) in Afghanistan. First and Third Platoons were assigned to Camp Mike Spann in Balkh Province in the northern region of the country while Second Platoon was assigned to Herat at Camp Arena.

The mission set was simple: provide security for transition teams in our AOs when they conducted site surveys or key leader engagements. Other aspects of the mission included guardian angel missions were Paratroopers guarded other contractors or Soldiers while off NATO bases or providing a Quick Reaction Force (QRF) for the base.

This deployment was vastly different than the Kandahar Province mission of 2012 and reflected the transition from NATO led counter-insurgency operations to an advise, assist, and accompany mindset that enabled the Afghanistan Government to take lead on running their government. In short, this wasn't an operation where we conducted offensive operations to clear villages of Taliban insurgents but to enable the host country after the United States had dumped billions of dollars into standing up a full land army in Afghanistan.

In theory, this meant the mission would be much safer, but the Paratroopers were not taking any chances, especially this one.

CPT Robinson designated me, SSG Dreeson, and a team leader, SGT Helbing as the company advanced party (ADVON) for the deployment along with the unit Supply Sergeant SGT Aljamaine Smith. We would be responsible for flying about ten days prior to the main body of Bravo Company to Afghanistan and setting conditions for the unit to conduct operations.

We had three main tasks: inventory the theater provided equipment (TPE), learn about the AO and what it entailed, and understand the dynamics of our mission and day to day life at Camp Mike Spann. 2LT Ziller and two of his Paratroopers conducted their own ADVON to Herat Province where his platoon would be based out of. Our flight would depart from Fort Bragg's Green Ramp on the last day of November in 2013.

Thanksgiving was celebrated at Fayetteville among friends that year. During this time in our lives, Michelle and I were not the best chefs and preparing a full turkey dinner was just not in the cards. Luckily, friends who had become like family to us invited us in. SSG Eric Myers and his wife Laura invited us to Thanksgiving, which we graciously accepted while MAJ Marcus Franzen and his wife Michelle also invited us to dine with them on Fort Bragg the next day.

Marcus had facilitated my interest in applying to become a Green Beret and submit a packet to attend the Special Forces Assessment and Selection Course for my year group. As the Special Operations Recruiting Battalion Executive Officer, his assistance was greatly appreciated; especially for a one-legged Paratrooper who needed to submit additional documentation proving he was medically qualified and capable.

Finally, on November 29, 2013, Michelle drove me to the 2Fury footprint for my departure. Understandably, she was concerned, and her anxiety showed. The last time I deployed, I almost died and returned missing a leg. What would happen this time?

She remained strong however, as she always had. After embracing and kissing her, I bid her farewell as I entered the battalion headquarters classroom with the other Paratroopers flying out on ADVON. As she drove off in my Dodge Ram, I hated knowing the next eight months away from her meant she was by herself in Fayetteville. I had nothing to fear though, this deployment is where Michelle truly earned her spurs as a quintessential Army wife.

The Battalion XO, MAJ McDermott was the Officer in Charge (OIC) for the ADVON Party. He ensured all personnel (PAX) were accounted for and that we had our weapons and carry on equipment for the flight. SSG Dreeson and SGT Helbing looked ready to and in good spirits. As the white school buses picked up speed towards Green Ramp, we had one last look at the fishbowl and footprint for the 4th Brigade Combat Team. As the Army was downsizing during this era, the brigade was to be disbanded and its battalions and squadron distributed throughout the remaining three brigades of the 82nd Airborne throughout the following year. Consequently, the new battalion footprint was to be relocated to a location unknown to me at this time. A new chapter of our unit would begin in 2014 after completing the mission assigned to us in the present.

Unlike my first deployment, there was no first-class seating. Across the length of the contacted Omni Airlines 737, all seating configurations remained the same. Regardless of rank or position, all Service Members were equals.

The flight first stopped at Cherry Point, North Carolina to pick up a detachment of Marines before lifting off for refueling at Shannon, Ireland. Our first layover was for eight hours. The OIC of the Marines on board told his fellow Devil Dogs to enjoy two beers and to not embarrass the Corps.

For the 82nd Paratroopers, things were a little different. MAJ McDermott asked the Battalion Operations Sergeants Major if the ADVON team could also imbibe. A soft but stern "no" was his answer. Even though the ADVON party consisted mainly of NCOs and Officers, the culture of the All Americans permeated this aircraft. We were officially deployed, and although General Order Number One for the theater did not technically apply, we were made sure to understand that this was all business. So much for a Guiness in Ireland.

For three days, the ADVON party waited in the Air Force Base at Manas near the capital of Kyrgyzstan, Bishkek. Apart from theater in-briefs and moving our baggage from building to building, we waited. The weather was wet and cold, with mud and puddles everywhere. We stayed in the giant circus tent structures trying to keep warm when not venturing off to the DFAC or the MWR to attempt to call home to our families. There was a fitness center though, and I strove to improve my physical fitness through strength training.

On the final day of our purgatory, we boarded the C-17. Taking off from the gateway to Afghanistan, I mentally prepared myself for my first return to this land since being wounded the previous year.

Bagram Airfield (BAF) is a sprawling complex located near the capital city of Kabul that serves as a major port of arrival and departure for coalition forces in Afghanistan. Upon landing, we were greeted by the departing unit, a battalion from the Guam Army National Guard. The 2Fury ADVON party was separated based off their final destinations in Afghanistan.

Most of the party headed to their base located adjacent to Kabul International Airport while the Bravo ADVON team went to either the western province of Herat or Mazer E Sharif's Camp Marmal to the north. The layover at BAF was brief, a smaller prop aircraft of dubious origins carried us over the Hindu Kush towards Balkh Province in the northernmost reaches of the country.

Mazer E Sharif had been one of the last cities to fall to the Taliban in 1998, but the first to be liberated by U.S. and Northern Alliance forces in 2001. Compared to my brief experience the previous year in the Pashtun dominated Kandahar Province, Mazer E Sharif contained many ethnicities to include Pashtun, Hazara, Uzbek, Tajik, and Farsi. In addition, it was much more progressive than the south, with women walking without wearing the Burka, girls attending schools, lively music in the streets, and billboards advertising everything from wedding halls to cellular devices.

The NATO base adjacent to the city was run and managed primarily by the German Bundeswehr and additional forces from throughout the Scandinavian countries.

Although our destination was Camp Mike Spann, about a thirty-minute drive from Camp Marmal, we were delayed due to weather conditions on getting a helicopter flight out. Contenting ourselves on exploring the large European base, myself, SSG Dreeson, and SGT Helbing marveled at the amenities and infrastructure. Multiple dining facilities, a bazaar, barber shops, an enormous MWR, duty free shops, and even a massage parlor were scattered throughout the base. It seemed each NATO contingent added their own piece of the home country to the base. Compared to the COPs we had been on in the past, Camp Marmal seemed like paradise.

After two days of wandering and waiting for a flight from a contracted company called Molson Air, we finally received word that a convoy from the company of Guam NG Soldiers based out of Camp Mike Spann was going to pick us up. We had only a finite amount of time of around two weeks to complete the ADVON process and so far, four days had already passed with transit to theater and waiting on pickup.

At the passenger terminal, a convoy of four tactical vehicles waited for us. A mix of MATVs and MAXXPROs, the vehicles looked to be in much better condition than the MRAPs we had in 2012. From what I could immediately discern, no scars of combat were present. The PL and PSG were from the First Platoon, the same PLT I would be conducting the relief in place (RIP) with. These leaders immediately warmed up to our presence, giving away to excitement that their journey was almost over.

After a brief series of handshakes and introductions, I boarded a MAXXPRO for the short ride to Camp Mike Spann. Sitting in the back, my ability to ascertain what the environment was like was severely limited. One thing I did notice though, was that when we departed the gate of Camp Marmal, the gunner in the turret did not rack a round in the chamber of his M240-B. In fact, there were no indications or radio traffic ordering weapons statuses to go red, something that was second nature to us Paratroopers in a combat environment.

I could sense the vehicles slowing down as they approached the Afghan National Army Corps base of Camp Shaheen which partially enveloped Camp Mike Spann. Pulling through the gates, they made a slow right turn into the NATO base which was to serve as my new home for the next six months or until the camp was turned over to the Afghan Army's 209th Corps. The ramp lowered to reveal the convoy situated in front of the camp's main entry control point at a series of clearing barrels.

Soldiers offloaded from the vehicles and lined up behind a clearing barrel to unload and clear their weapons. Almost all of them I noticed, had never loaded a single round into their personal weapons or even bothered to load a magazine.

The outgoing unit graciously escorted me, SSG Dreeson, and SGT Helbing to a transient barrack to unload our bags prior to giving us a tour of Camp Mike Spann. The Soldiers on this base did not live in tents but in wooden barracks called bee huts. These buildings were rectangular in shape, with a hallway running down the middle in its entirety. On each side of the hallway were rooms separated by plywood from floor to ceiling. Each room was about six feet by five feet where one occupant resided. This was a major step up compared to the spartan living conditions we had endured previously. Immediately after dropping our bags and taking off our armor or kit, we met up with the platoon leadership for the base tour.

Two platoons of one U.S. Army Company lived on the base. The Third Platoon was in Herat where 1LT Ziller was conducting his own mission. A short walk from the bee huts stood the platoon office, a concrete building divided into two quadrants; one room with desks and non-classified computers for the platoon leadership and Squad Leaders, and a larger room which served as an ad hoc meeting area for the rest of the platoon.

Adjacent to the First Platoon suite was the Company Headquarters where CPT Robinson, 1SG Goodart, and the Company XO, 1LT Laroque would reside. Further down an avenue of more bee huts, we arrived at the Company Tactical Operations Center (TOC) where missions were tracked. This small building would house the Bravo Company field artillery Paratroopers that were attached as forward observers, a Fire Support NCO, and the Fires Support Officer. Unfortunately, there were no artillery or fires missions in Regional Support Command – North so their duties naturally were used in tracking Paratrooper movements and providing a Counter Insurgent Support Team (COIST) that worked to provide intelligence of the operational environment to the Commander and the Paratroopers conducting missions throughout the region.

We also visited the Third Platoon suite where 1LT Kamal Wheeler's team would reside. The setup I noticed was a lot nicer than mine and I would be lying if I didn't contemplate requesting a switch up. Their facility had desks for all leaders but their platoon staging area was twice as big as First Platoon's and had cubbies for all personnel to stage their kit for missions.

The Camp Mike Spann Quick Reaction Force (QRF) was staged from this room, a tasking that switched every couple of weeks for both platoons. The company motor pool was divided into two areas. First Platoon on one side of the base and Third on the other. This meant my guys had to reposition their vehicles to other staging areas prior to each week on QRF.

The tour also covered the latrine and shower facilities, the gym, and the laundry facility with a two-day turnaround. The DFAC I noticed was exceptional with a plethora of options for all of the base's residents. Since other nations of the coalition inhabited the base, a variety of meals were available.

After viewing the Basic Aid Station or BAS, we were turned loose for the evening. SSG Dreeson, SGT Helbing, and I planned what our priorities were in accordance with CPT Robinson's intent for our AD-VON mission. I would go over the property books with the platoons to ensure all equipment that the boss was signing for was present and fully mission capable while the NCOs learned about the battle rhythm and the nuances of the missions our platoon would undertake for this tour.

For the next ten days, our battle rhythm focused mainly on the inner workings of Camp Mike Spann and observing the outgoing Soldiers duties and responsibilities. This observation was skewed however, as these Soldiers were primarily focused on redeploying back to Guam and their demeanor showed it. That's not to say they weren't focused. Far from it, in fact.

The Soldiers' patrol briefs were some of the most detailed I had ever witnessed, complete with rehearsals and even an invocation prayer to bless the mission. This was one of the tightest units I have ever worked with, and their love of duty and Guam showed.

Despite the appearance, there were some glaring issues. For starters, there seemed to be no acknowledgement of a threat in the AO. Although weapons were mounted to tactical vehicles moving out to perform Security Force (SECFOR) missions, I never once observed weapons statuses changing to reflect a threat. The unit had never been engaged or encountered a threat during their missions, despite a member of their battalion being killed previously on the deployment in a different province.

Complacency kills and I worried not just for their safety, but for mine as well when I rode along observing the area of operations. Regardless of their posture, I know for certain that SSG Dreeson, SGT Helbing, and I were on a red status or had a round in the chamber of our primary and alternate weapons whenever we left the base for a patrol.

In addition, the property of the unit was in complete disarray. Sometime after finalizing the property in Herat for Second Platoon, SGT Aljamaine Smith, the Company Supply Sergeant, arrived by Mosul Air helicopter to Camp Mike Spann to facilitate the property inventories. There was a complete absence of shortage annexes for the theater provided equipment (TPE) so we had no basis besides a technical manual (TM) of what components and basic issue items came with a particular piece of equipment.

When equipment was laid out for our inspection, it was just the end item. For example, a lot of items such as a 153 Common Remotely Operated Weapons System (CROWS) have multiple components that are a part of the system. Each item is listed on the shortage annex so that the end user can see if all pieces of equipment that make the system work are in fact present and functional. For many pieces of equipment, there was no document so we had to painstakingly recreate the shortage annexes from a TM so that CPT Robinson could sign for the equipment with confidence.

Soon, the main bodies started arriving in theater on two flights. We accompanied the Guam Soldiers to Camp Marmal to pick up the Paratroopers and their accompanying baggage. SSG Dreeson and SGT Helbing already worked with the outgoing NCOs on getting the billeting ready and for the team to immediately begin their left seat / right seat with their counterparts.

CPT Robinson, 1LT Laroque, and 1SG Goodart arrived on one of the first flights and immediately begin immersing themselves into the command suite to begin the process of taking over SECFOR operations and guardian angel missions from the redeploying Soldiers. It was a busy time, as most transitions during deployments are, with no shortage of tasks and patrols that still needed to be conducted, despite our arrival.

SFC Petrik and the men of First Platoon set about getting things in order as 1SG Goodart advised. Although the Soldiers from Guam did things their way, the 82nd Airborne's culture had their own way as well.

For starters, the Paratroopers, given their basic combat loads for the M4 would always be amber on the base and red off the base. This meant a magazine in the well of the M4 with no round in the chamber when on friendly turf, and one round in the chamber when on a mission. The reasoning was that leaders wanted all Paratroopers to know that there was a threat in our AO, and we always were to be ready to react to any situation.

Unlike the Guam Soldiers on the base, we had no female Paratroopers in our company, so that was one less dynamic for the 1SG to focus on. Enlisted were assigned to their bee huts by their platoon and squad to make mission command that much easier in case of emergencies or training drills. The machine guns on top of the tactical vehicles were all prepped every time we left the gate and battle drills were rehearsed such as reacting to near ambush or contact. While we conducted left seat right seat, we patiently observed the outgoing Soldiers, took notes, and then proceeded to take over operations. This didn't happen overnight, but within a weeklong process.

Our final validation was a SECFOR patrol to an Afghan National Army Base about three hours south of Camp Mike Spann into Samangan Province. The Regional Support Commander – North (RSC-N), COL Walter Sweetser, was the validator for our patrol. In our MAXX-PROs and MATVs we drove throughout Balkh Province passing Mazer E Sharif and the Blue Mosque, one of the most holy sites in Islam and through the scenic Khulm Pass into Samangan Province.

Radio chatter was kept to a minimum with gunners scanning their sectors for threats. If we needed to make an adjustment, we did it over the net. Overwise, complete silence except for the buzz of the counter-remote control IED jamming systems known as "Dukes" and the sounds of the MRAP engines.

After a quick stop for relieving ourselves at our destination, we turned around and returned to base the exact same way as we came. After we pulled through the gate, cleared weapons, refueled, and downloaded weapons and gear, we conducted our first After Action Review (AAR) of the mission. Besides the token comment to communicate more throughout the mission (maybe he was bored), there were no other major injects from COL Sweetser. We were now validated to assume SECFOR patrols in the region to support RSC-N.

The Transfer of Responsibility Ceremony took place on a sunny day on the base where CPT Robinson assumed responsibility for the mission. The Guam Soldiers finally departed for their homeland in mid-December, leaving us to our own devices. They accomplished their mission with pride and honor, and nobody could ever take that away from them. Apart from a property fiasco where multiple pieces of TPE sensitive items were unaccounted for (they were later found or acquired thanks to our supply team), the relief in place was a success.

The battle rhythm for Bravo Company henceforth was as follows: on one week, First Platoon took SECFOR missions escorting Army Reserve engineers to conduct site surveys and Key Leader Engagements at local Afghan military or government facilities. When not on patrol, we trained in dismounted marksmanship behind Camp Shaheen, conducted vehicle rehearsals such as recovery or gunnery drills, or simply planned for the next mission.

Third Platoon's week consisted of conducting guardian angel missions. This entailed escorting U.S. or NATO personnel onto Camp Shaheen as bodyguards to thwart potential green on blue attacks and standing up the base QRF. It wasn't too busy nor was it ever dull for us. We were not responsible for maintaining force protection (FORCEPRO) of the base which included manning guard towers or the ECP. That was the Armenian contingent's responsibility, and we were thankful not to have that tasking.

For the remainder of December, the unit quickly settled into its new routine. When not conducting planning, executing missions, or attending briefs, I typically sought self-improvement in some form or capacity. One of the ways many Soldiers achieve self-improvement on deployments is through strength training.

Some bases allow for running on designated routes or improved trails, depending on their size. Camp Mike Spann had a small trail running along the inner wall of the perimeter that many of the B CO Paratroopers utilized. From 0500-0630 in the morning, I worked out at the fitness center just a short distance from my bee hut. Religiously, I lifted weights and ran about three to four miles a day on a treadmill to start my days. Preparing for SFAS was in the forefront of my mind and both 1LT Ziller and I had submitted our packets to attend the course hopefully in 2015.

Breakfast was not terrible on the base. While spartan when compared to the massive DFAC on Camp Marmal, the Camp Mike Spann DFAC still provided plenty of variety for all palates. Thus, nobody ever went hungry during this deployment and MREs were not a regularly consumed meal for the most part. ECOLOG and KBR provided the contractor support for the DFAC, which meant our Paratroopers were not tasked to serve meals or clean up in the kitchen or dining hall.

The MWR on the base did an exceptional job with the resources it had keeping all units entertained. Poker tournaments were put on which I took part regularly (no cash or gambling in this case, just bragging rights).

CPT Robinson regularly participated in the basketball tournaments with members of the company participating. It quickly became apparent that the new kids on the block from Fort Bragg did not mess around in competitions. The number of trophies and titles claimed quickly stacked up. The United Services Organization (USO) even brought to Afghanistan and our base professional Mixed Marshall Artists who sparred with many of the Paratroopers, to include myself. Getting my ass handled to me by these guys was a unique workout and method to burn off steam.

Christmas was celebrated with a day off from patrols. Despite the amenities offered to us, we were still deployed thousands of miles from our families. It is typically a hard time for all involved during this time of the year with feelings of loneliness for many, especially those that were single.

SFC Petrik came up with a unique way to bring the platoon together that morning since we had no missions that day. A Texas Hold'em tournament was put on in the First Platoon suite. The winner received a new set of headphones and there was also a nice prize for the runner up. I do not recall playing in this one though. As a leader, it was a lose-lose scenario if I won so I spent much of the day talking to Michelle while she celebrated Christmas back home with her family in Kentucky.

The DFAC pulled out the stops with an extravagant Christmas feast that even had non-alcoholic champagne.

2014

The new year started with some new transitions. By this point, CPT Robinson was due to change command at the end of January with the incoming commander, CPT Bryan Blackburn taking the guidon and leading alongside of 1SG Goodart.

With the change of command came the numerous inventories of organic and TPE property that all had to be accounted for down to the 10mm socket in the mechanics tool set. 1LT Laroque led the efforts to prepare for the change of command. I professionally developed as we learned how to conduct layouts in accordance with the property book. Items were laid out by their order, grouped by type, and lined up in sequence by serial number to make the inspection run smoothly and efficiently.

Prior to this, I had just grouped everything together and my team frantically searched for the correct serial numbered piece of equipment. This technique is something I continued later throughout my career and am damn glad I realized this when I was a younger officer.

The days just prior to the change of command and CPT Robinson's final as the Bravo Company Commander were marked with comradery and team building events. Brutal Ball, a hybrid of football, ultimate Frisbee, and rugby was played in the motor pool between First and Third Platoons.

It was a perfect sunny day in late January with unusually warm weather for that time of the year in Northern Afghanistan. The bouts became heated between the platoons as the rivalry was strong. One of the Paratroopers, SPC Ryan Daley, with whom I had the privilege of reenlisting on a C-130 Hercules over Luzon Drop Zone prior to an airborne operation the previous fall, nearly came to blows with one of 1LT Wheeler's Paratroopers due to the intensity. Luckily, his Squad Leader, SGT Tippit and SFC Petrik resolved the conflict prior to anything getting serious.

And just like that, on the day of the change of command, a blizzard rolled into Balkh Province that dropped over a foot of snow overnight. The warm weather quickly dropped into the teens. Rehearsals for the ceremony were conducted in the mechanics bay where there was space and heating. LTC Zieseniss and CSM Kelly were on site to oversee the ceremony and enable the tradition of the outgoing commander handing over the guidon to the incoming commander. 1SG Goodart rehearsed the event to his satisfaction, ensuring that the two platoons rendering of the All-American Song was adequate. 1LT Ziller and Second Platoon remained in Herat Province due to their missions.

The ceremony was over quickly, I do not distinctly remember CPT Robinson's speech being long so I must have liked it. CPT Blackburn's speech was also short, the custom for incoming commanders. A new era for Bravo Company was initiated at this point. CPT Bryan Blackburn was now at the helm as Brutal 6, and he immediately began a transformation of Bravo Company that would have long lasting effects.

Josh reenlisting SPC Ryan Dayley on an airborne operation (NOV 2013)

CHAPTER 15
BRUTAL NATION
JANUARY TO AUGUST 2014
BALKH PROVINCE, AFGHANISTAN

The temperatures in February plummeted to a low of minus 25 degrees Fahrenheit, freezing everything. Engines on the MAXXPROs had to be started up to an hour prior to a mission to get them ready for the SECFOR missions. The First Platoon suite, made of concrete, was almost unbearable to work in. I remember wearing all of my extreme cold weather gear at my desk while shivering. The outdated heating failed to raise the temperatures in that office so, naturally, it was vacant most of the time.

On the heavy MAXXPROs, chains were attached to the tires to prevent sliding on the ice-covered roads within Balkh Province. At one point, while we were conducting marksmanship operations at the Camp Shaheen range complex, one of our MAXXPROs slid off the road and almost flipped. Luckily, my driver, SGT Pritchett, a prior Marine and a master behind the wheel, was able to negotiate the snowbank and safely return the MRAP onto the snow-covered route. However, a premature call for assistance had the Third Platoon QRF escort SSG Kenny, our mechanic and his wrecker out to the site to support us. They arrived just as SPC Pritchett overcame our dilemma. Even though they weren't needed, for a week at least I never heard the end of it from 1LT Wheeler about how his men came to our rescue.

Despite the weather, we continued our duties in escorting the engineers to the designated sites for inspections and surveys. The poor weather conditions however had one benefit, it kept most civilian traffic off the road. My gunner, SPC Lamay, who was quick to make a remark, always commented that his upper torso, exposed to the elements, was soon numb after beginning any patrol. Regardless of how cold it was, First Platoon and every other Paratrooper in Bravo Company carried on its mission and duties to the absolute standard.

CPT Blackburn wasted no time pulling 1SG Goodart in to discuss the vision of Bravo Company. After a combined effort, the Command Team published their expectations for the Paratroopers. Henceforth, Bravo Company was formally known as Brutal Company or Brutal Nation among the Paratroopers guarding its guidon. CPT Blackburn gave separate talks to junior Enlisted, Team Leaders, Squad Leaders, Platoon Leaders and Platoon Sergeants, and his XO 1LT Laroque to get a feel of the command climate and culture of the organization.

By separating leaders from subordinates, he was able to get candid feedback that empowered him as the Commander to facilitate changes that would make the unit that much more cohesive and ready. I would later utilize this communication technique myself.

With less than four months until we handed over Camp Mike Spann to the Afghan National Army and retrograded to Camp Marmal, planning efforts commenced to slowly tear down key facilities and infrastructure that the U.S. Department of Defense no longer needed to maintain. Traditionally, retrograding units are vulnerable without adequate security, so we wasted no time in perfecting our skills in precision marksmanship, combat lifesaving, small unit tactics, and command and control.

Base wide defense training exercises were conducted. 1LT Garcia, on the base intercom system, sounded indirect fire attacks with "IN-COMING, INCOMING, INCOMING," driving Paratroopers to occupy battle positions for an expected imminent attack. In one exercise, after many of the occupants had already vacated their quarters, the two platoons cleared almost half of the empty base, even exercising a mass casualty scenario in a former DFAC with German Bundeswehr MEDEVAC assets from Camp Marmal. In short, we were anything but complacent, even while the enemy was still hibernating.

Amid all the training, guardian angel, QRF, or SECFOR missions, one prospect was cherished by all, a trip to Camp Marmal. About once or twice a month, the engineers we escorted needed to venture to Camp Marmal for business. Lasting for almost eight hours when you factor in the 30-minute drive time each way, the trips to Camp Marmal filled every MAXXPRO seat with Brutal Paratroopers who wanted to get off Camp Mike Spann.

With its enormous MWR, multiple DFACs and Post Exchanges, and a bazaar with Pizza Hut, Green Beans Coffee, local shops with internet cards, and a massage parlor, it was a whole new world. Thus, when the convoy of giant MRAPs rolled into the parking area adjacent from the MWR and the engineers departed, SFC Petrik gave the boys a time to return, and the Paratroopers scattered.

I received multiple care packages from my family and Michelle throughout that deployment, so I did not lack for anything. Thus, my time was spent in the MWR where I signed in to occupy a small room to watch a film of my choosing. Dark, quiet, and with a couch all by myself, it was a mental break from the deployment and a rare opportunity for me to relax. I also walked to the DFAC with SFC Petrik for lunch. With its multiple lines, a salad bar, panini presses, and assortment of choices, it was a feast for the eyes.

When the platoon came back together in the afternoon to debark for Camp Mike Spann, boxes and bags took up whatever remaining space in the vehicles we had. Even the engineers apparently took the time to go shopping. As we drove through Mazer-e-Sharif and passed the famed Blue Mosque on our way home, I could tell that I would get through this deployment unscathed. I had only four months left and soon we would permanently take up residence at Camp Marmal, drastically improving our quality of life and eliminating another concern for our wellbeing.

Apart from missions and planning, additional duties occupied my time overseas. As an officer, it was an additional responsibility for me to facilitate the monthly inspections of the commander's property. Between 1LT Ziller, 1LT Garcia, 1LT Wheeler and myself, we took turns every month accounting for each piece of sensitive equipment and a ten percent inventory of all pieces of equipment's components. 1LT Laroque had a rotation in place, so we knew which month our duty was to conduct this process. In addition, the lucky winner had to travel to Herat to account for Second Platoon's equipment and attached company property. When it was Luke's turn, he traveled to Balkh Province.

In early spring, my number popped up. After a couple days of accounting for all property and looking everywhere for those onesies and twosies that always seem to misplace themselves, I traveled to Herat via Kabul International Airbase, the Headquarters of the NATO mission in Afghanistan.

Unfortunately, my flight from Kabul to Herat was delayed by almost a week so I was stuck in transient quarters. SFC Petrik had the platoon up north, so I was essentially unemployed for a week. That wasn't to say I didn't stay busy. Apart from the magnificent dining facilities that the senior leaders had, I spent some time at the gym or wandering about. COL Watson, the Brigade Commander, took me on a tour of the headquarters when we bumped into each other. The complex was huge, and I remember the giant screens showing the common operating pictures, weather, significant activity, and MEDEVAC status for the entire theater.

It was an eye-opening moment to see the heartbeat and brains of Operation Enduring Freedom Afghanistan. At some point, my wounding and subsequent MEDEVAC must have been on one of those screens I now realize, and without those staff officers and NCOs monitoring and reporting, I could have been worse for wear. Those Soldiers truly don't get enough credit but surely drank enough coffee for their efforts while assigned to that cavernous theater like room.

Finally, after hours spent watching films in the transient barracks, I boarded an Italian C-130 Hercules to Camp Arena in Herat Province. I don't get air sick often but on the decent, I felt a sudden urge to heave as the aircraft conducted a combat landing on the dusty tarmac. Luckily, the light breakfast I had prevented any unfortunate messes for my Italian hosts during that flight. Located a short distance from the Iranian border, Herat looked more like a desert environment when compared to the mountainous regions of Eastern and Northern Afghanistan.

1LT Luke Ziller and SFC Willie Wentworth III greeted me at the PAX shed with open arms. As usual, Luke had his lightheaded and witty demeanor while his Platoon Sergeant had an unimpressed attitude that occasionally revealed a grin. I had known Luke since 2012 and we executed GORUCK events together the previous summer. Willie and I had the pleasure of attending Air Assault School at Fort Benning, Georgia back in 2011. He was at that time a Ranger Instructor and me, a fledgling Ranger graduate and novice 2LT. After throwing my ruck into a non-tactical vehicle (NTV), we sped away from the airfield to their living area on Camp Arena.

Camp Arena was primarily occupied by Italian and Spanish military contingents with a small force of U.S. Soldiers. Luke reported directly to an O-6 at the base and his Second Platoon basically represented the predominate U.S. presence on site. Their accommodations consisted of air-conditioned Force Provider tents for their Paratroopers, much more spartan than what the rest of Brutal Nation resided in. Their DFAC too lacked many of the meals Americans enjoyed, catering mainly to the Spanish and Italian forces.

I distinctly remember previously how SFC Wentworth commented that the fish in the DFAC still had their eyeballs when served. Regardless though, Luke had a degree of autonomy for him and his men that would be the envy of others. His operations tempo was much more strenuous than mine or 1LT Wheelers. Unlike First and Third Platoon, who could switch on SECFOR patrols after every week or two, Second Platoon patrolled weekly and sometimes daily.

The inspections were conducted in less than two days, even though I was on ground for almost four. I observed their operations and 2LT Ziller's mission command post where his team tracked patrols. It was a solid setup and seemed to function efficiently. CPT Robinson and 1SG Goodart made the right choice in assigning the Regional Support Command - West mission to Second Platoon. Some of the mental notes I made during this visit had later implications when I was forward deployed as an independent unit to Iraq in 2017.

On my last night in Herat, Luke and Willie hosted me at the Italian pizza joint on the base. It was delicious and perhaps one of the deployment's highlights as it would be sometime before I saw the Paratroopers of Second Platoon again. On the way back to Balkh Province, I had a much shorter layover at Kabul International Airport. A small prop-engine aircraft returned me to Camp Marmal from where I took the Canadian Molson Air helicopter back to Camp Mike Spann. SFC Petrik and the boys held the fort down well, so the almost two-week trip didn't leave me with a ton of issues needing to be addressed.

Meanwhile, Brutal Nation was blossoming into a full-fledged culture. Hats, stickers, and apparel were designed and made by the Command Team for the Paratroopers. The Kryptek Spartan logo was incorporated into the Brutal Nation design, but I doubt the Kryptek Organization was ever aware of the copyright infringement nor do I believe they would have made a huge fuss about it as our organization wasn't financially gaining from it. From the Facebook page to our offices, and all of our email signature blocks, Brutal Nation was everywhere, and we wanted everyone to know it. It wasn't just our unit that was experiencing changes, but our mission was too.

Usually, on long distance patrols, we drove for hours in our MRAPs. However, CPT Blackburn and 1LT Laroque made inroads with the aviation unit at Camp Marmal since the change of command. This meant we began conducting air insertions from UH-60 Blackhawks on long range patrols. The engineers were too glad to spend less time on the road and for security reasons, we were too as the weather was starting to warm up; the signal for the upcoming fighting season in Afghanistan to begin. Prior to one air infill, the entire platoon shaved their heads for the mission. It was not forced or coerced but 100% of the platoon magically decided to shave their heads in comradery. Perhaps to match the Company Commander's hairstyle.

Flying above RSC-N was a joyful experience as the crew chiefs often allowed the doors to remain open, letting in a nice breeze and allowing everyone to enjoy the spectacular scenery of the Marmal Mountains just south of Maser-E-Sharif. We would touch down, secure the engineers, allow them the freedom of movement they needed, and twenty minutes prior to the survey or meeting's completion, SGT Demetrius Dasher our FO, called the birds back in. We boarded and took off for Camp Mike Spann. These were the first operations involving air lift that I had conducted in theater and to this day, I prefer flying than driving when the threat of enemy air defenses is absent or negligible.

That spring in Afghanistan passed quickly and without much incident. As the base progressively emptied of NATO forces, our living conditions became more austere. A formal Transfer of Authority (TOA) ceremony was scheduled to be held on April 28, 2014, on Camp Mike Spann between representatives of the Afghanistan Government and NATO to herald the handoff of the base to the Afghanistan National Army's 209th Corps based in the neighboring Camp Shaheen.

Prior to that date, U.S. and multinational contingents divested the camp of unit and national property. The Afghan military received infrastructure and some logistics for the base, but the military equipment was to be retained. Even the symbol of our predominate religion in the west, Christianity, was dismantled when the chapel's steeple was removed as to not offend the new tenants.

On our last month on the base, some new experiences were encountered. Third Platoon toured the adjacent Qala-i-Jangi fortress where the uprising of Taliban prisoners on November 25, 2001, killed the Central Intelligence Agency's agent Mike Spann, the namesake of our current home. From all accounts, it was a sobering and important reminder of why we were there in this country. As First Platoon was running SEC-FOR patrols that week, we were too occupied to attend.

A couple weeks prior to the handoff, our XO, 1LT Laroque, was sent by CPT Blackburn to Camp Marmal to begin preparations for our arrival. He worked feverishly to establish a footprint for Bravo Company which consisted of four large Force Provider tents, two for the platoons, one for the Tactical Operations Center, and one for the command team. Barracks were identified and we even had a small fleet of NTV and John Deere UTV Gators to assist us in getting around on the sprawling base of Camp Marmal. Due to his efforts, Bravo Company would not have a loss of capabilities during the retrograde and transition.

As occupants vacated, contractors also begin departing. Luckily, it seemed that laundry, showers, and chow were among the last to end. When the DFAC finally shuttered operations, we dined on military hot rations or MERMITES, which are bulk entrees and sides served in preheated packages. Unfortunately, I developed a severe allergic reaction to the preservatives in the food, causing my back to develop painful rashes that lasted well into the summer. The equipment in the gym was also destined to leave. Weirdly enough, it was on no property book and thus became spoils for the unit. Bumper plates, barbells, dumbbells, and squat racks clandestinely found a home in our organic containers that would be shipped back to Fort Bragg. I bet my right leg that some of that equipment continued the journey further to the homes of some of the most dedicated lifters and cross fitters in the Brutal Nation family.

Shortly before the Transfer of Authority ceremony on late April, I received a new Platoon Sergeant and Weapons Squad Leader. SFC Petrik switched out with SSG Christian Lawson, a recent Ranger School graduate and former Squad Leader in Third Platoon. SFC Petrik moved to the Battalion Operations Cell in Kabul to continue his work while SSG Lawson stood ready to lead First Platoon. A Kentucky native and fellow outdoor enthusiast, I had no trouble getting along with this hard charging NCO who would become one of my most trusted confidants. SSG Christian Nooney, a reputable NCO who had previously served with 1SG Goodart, also joined the team. First Platoon had three squads now to finish the rotation and the additional experience that SSG Nooney provided was gratefully welcome.

Finally, on April 28, 2014, the closure of Camp Mike Spann was made official. Representatives from NATO and the Afghanistan National Army, as well as media outlets, arrived for the ceremony. Due to the increased security risks from insider attacks, Bravo Company Paratroopers were tasked with not only providing a complement of Paratroopers in a formation to represent the American contingent but to ensure security was provided. Our designated marksman, obscured from view, scanned the ceremony and the surrounding environs of any perceived or potential threats. The base in essence, was locked down tighter than it had ever been during our tenure. Thankfully, the ceremony passed without incident.

Order of march for the relocation to Camp Marmal was the Brutal Company Headquarters, First Platoon, and then Third Platoon. Prior to pulling out of Camp Mike Spann I did a solo tour of the ghost camp. Remnants of units prior to us still left their marks. Former outfits from the Guam and Alabama National Guard units who preceded us were prevalent everywhere. This base, around for over a decade, had seen a lot of growth and transitions as units occupied and departed.

As the last force from NATO to reside here, it was interesting to me during this time how the Afghan forces would take over this place. As I climbed into the MRAP for the convoy out, I took one last look at the base, my home for the last six months, and loaded a round into the chamber of my personal weapons system. SGT Pritchett, my driver, gave me the thumbs up, his custom every time I completed this ritual.

Camp Marmal for Brutal Company took a lot off our plate and provided us with opportunities that had been, until our arrival, limited to occasional visits. Perhaps most welcoming for us was the dining opportunities. Breakfasts, lunch, and dinners at Camp Mike Spann paled in comparison and if we didn't watch ourselves, we could be slightly heavier under a parachute canopy upon our return to Bragg.

To keep the boys in shape, we did regular PT and even conducted the Army Physical Fitness Test (APFT) for Paratroopers who needed to promote or attend Army courses back home. Our work areas were also improvements, with our vehicles being located only a dozen meters from our offices. A couch and large flat screen television were even in our tent playing the Armed Forces Network which was a huge win for everyone.

Once Brutal Nation established itself at Camp Marmal, a final change in personnel was conducted. 1LT Mike Laroque was promoted to Captain and transferred to the Battalion Operations Cell in Kabul. His replacement was 1LT Douglas Ausenbaugh, another respected and immensely popular officer in 2-Fury who had the nickname "Ausen-Bro" due to his relaxed and easy demeanor. Mike, by virtue of his branch, Armor, was an outsider in an airborne infantry battalion, but his leadership and competence proved he was anything but incapable of belonging. Mike had deployed with the battalion back in 2012 as a Platoon Leader and had left his mark in Bravo Company. I haven't seen him since his departure, but I am sure he went on to have a highly successful career in the U.S. Army.

Our mission evolved as well. While we no longer maintained the QRF or conducted as many guardian angel operations, we were tasked with securing a new installation on an Afghanistan National Army Base. The Operation Coordination Center – Regional (OCC-R) was in downtown Mazer-E-Sharif and consisted of one building in the middle of the base. Its purpose was to enhance security, provide situational awareness, and report and share intelligence across NATO and our Afghan hosts.

Occupied currently by Norwegian and Finnish contingents also based at Camp Marmal, we would be tasked with securing the location until its closure that summer. A barbed wire chain link fence was all that separated us from our Afghan neighbors and due to the company's history of green on blue attacks on recent deployments, no chances were being taken with security.

The rotation was as follows: one platoon secured it a week and then switched out. The platoon not guarding the facility was conducting SECFOR missions, as was our usual custom. Only two squads were needed at the OCC-R with leadership being the Platoon Sergeant or Platoon Leader who generally rotated out every two to three days.

As the first rotation, our priorities of work were to improve upon the security situation. SSG Lawson and the Squad Leaders developed sector sketches of the whole compound and range cards for the crew served weapons that guarded the avenues of approach to the OCC-R. A small command post was developed in a vacant room with multiple monitors showing the closed-circuit cameras watching our surroundings. It wasn't comfortable but at least it had heating and, more importantly, air conditioning as the approach of summer was coming fast.

Food was MREs or whatever Paratroopers brought with them from the DFAC or PX on Camp Marmal. It wasn't a glamorous assignment by any means, but it did provide our boys with downtime to relax when not standing watch or conducting overwatch patrols on the base, something nobody complained about because it meant getting out of the structure to stretch our feet.

CPT Blackburn had us drill contingencies for the OCC-R to include aerial MEDEVAC, resupply, or ground evacuation in the event the location became untenable. The training was necessary for all parties involved to understand what was at stake if any enemy elements decided to attack in force. We were confident in the aviation assets to come to our assistance because they never failed to respond. We were more confident in each other though, after months, and in some case years, of training alongside one another and enduring conditions of various hardship.

Earlier on the deployment, I observed a flyer at Camp Mike Spann for a marathon race to be conducted at Camp Marmal on Memorial Day. Dubbed the Marathon of the North, it comprised of two 13.1-mile loops throughout the giant NATO base. I had never run a marathon before and decided to give it a go. Two other First Platoon Paratroopers, SGT Nunez and CPL Epps also wanted to attempt it. I mean, how many people can say they have run a marathon before, let alone in a war zone like Afghanistan.

My training consisted of four mile runs on the treadmill in the cold winter months while CPL Epps, a natural runner conducted regular five mile runs on the on-base trail. From December to May, I lifted and ran in the mornings or free evenings and soon accumulated over 500 miles ran during the first five months of the deployment.

On race day, in our Army Physical Fitness Uniform (APFU), we ensured we ate a small breakfast and had our hydration goo's and gels. I had absolutely no clue how to execute this, so I stocked up on as many energy sources as possible. There was also another factor, our OCC-R rotation began that afternoon, a variable which meant that not only did we have to finish the race and kit up for a two-day trip, but there would also be no time for a trip to the aid station if my leg started hurting. I had to be smart about this. I was a Platoon Leader first and being unable to leave for a mission due to a race would have been strongly frowned upon. Luckily though, I just had to get to the OCC-R, where I could sit without my prosthetic.

The three of us ran together for the first half of the race. I felt great for those 13 miles, having trained for a race in my spare time nearly every day. As the miles melted away though, so did my feeling of invincibility as hotspots on my right foot and inside my Flex Cheetah socket developed. For almost four hours, the three of us pressed on, our quest to run a marathon in Afghanistan in honor of our fallen as our only thought.

Finally, after four hours and some change, I crossed the finish line. There was no time for accolades or pats on the back as we needed to kit up for the patrol. Once the adrenaline wore off, I could tell I was going to be sore for the next few days. And indeed, I was. Nonetheless, I ran the first marathon of my life. I did it with one leg. And I did it in Afghanistan. The body is capable of some truly amazing feats when the mind is disciplined and focused. Later, I put in SGT Nunez and CPL Epps for Army Achievement Medals for their training regimen and accomplishment. CPT Blackburn even pinned one on my chest. What a hell of a way to earn a medal in Afghanistan.

It was during our tenure at the OCC-R that I truly began to get to know my men. Since we were cramped up into a single building with one floor and about eight rooms, we had to eat, sleep, and work together, regardless of rank or position. First Platoon truly represented the two A's of the 82nd Airborne Division patch on our shoulders. Standing for "All-Americans" because the Soldiers of the First World War came from all states and territories, my platoon also had a wide spread of men from the United States.

For starters, we had farm boys, like SPC Maglott who we often joked came from an Amish family but ran away to fight. We had SPC Mexicotte whose wife was a dedicated volunteer for the Bravo Company Family Readiness Group (FRG) and had the most kids out of all of the men. We had the gym rats, such as SGT Allen and SPC Daley. And we had the son of an acclaimed songwriter, producer, and director too. SPC Ian Harrison, the son of John Harrison, was one of the most humble but intelligent men in the platoon. We talked at great length about films and music, and I once commented that if I ever wrote about our experiences, I'd send the story to his father, though I doubt it would ever top Dune.

Not long after we began babysitting the OCC-R, Third Platoon was called up on a unique SECFOR mission. The Kholm Pass, a narrow canyon between Balkh and Samangan province experienced an earthquake. The highway running through the gorge had collapsed into a river below, thus blocking all land traffic along that route. Engineers were called up to survey the extent of the damage and Third Platoon happened to be on call at just the right time. Kamal Wheeler and his team were air lifted just to the north of the pass and provided overwatch during the engineer's inspection. As with all our prior patrols, no enemy action occurred, and the mission went off without incident. However, First Platoon was about to have their moment of fun as well. This time, it wasn't for a one-day sightseeing trip but for an extended period supporting units that were going kinetic every day and night.

Josh's platoon in Afghanistan after a mission, (MAR, 2014)

LEFT: Josh and his Platoon Sergeant, SFC Ray Petrik, (FEB 2014)
RIGHT: Prior to a mission in Afghanistan, (APR 2014)

CHAPTER 16
KUNDUZ
JUNE TO AUGUST 2014
REGIONAL SUPPORT COMMAND - NORTH, AFGHANISTAN

In June, Operational Detachment Alpha's (ODA's) of the 5th Special Forces Group under Combined Joint Special Operations Task Force – Operation Enduring Freedom (CJSOTF-OEF) were conducting missions with their Afghan Commando counterparts in vicinity of Kunduz Province to our east. CPT Blackburn was solicited by their elements to provide QRF support for the teams as they assisted their Afghan partner forces in combating Taliban forces around Kunduz. Naturally, our CO knew Brutal Nation would only be happy to support. Even better was that this time, First Platoon was on SECFOR and would be on deck to support this new kind of mission.

The Blackhawks touched down at a former NATO base that had been turned over to the Afghanistan National Army the previous year. My counterpart from Guam had his platoon stationed out at this base for much of his deployment. But for now, it was Brutal Nation's turn. As the crew chiefs pushed the doors back, eleven Paratroopers from First Platoon hit the deck from Chalk One, throwing their rucks out front and sprawling behind them. We scanned our immediate surroundings as the bird lifted off, the base looking deserted and a bit worse for wear. Picking up my ruck, I moved to the base of the landing pad and watched the remaining birds drop off the remainder of the platoon and Captain Blackburn. After consolidating, the squads moved off to predetermined locations to establish security around the perimeter of the former NATO base.

SSG Dreeson took his squad to a wall and some buildings near the fueling trucks positioned at the Forward Arming and Refueling Point (FARP) for the CJSOTF birds. He would be securing the FARP and the compound wall adjacent to it. SSG Nooney took his boys immediately to the building and tower near an unused entry control point. SSG Tippit moved into an adjacent compound and occupied a multi-tiered set of containers. With barrels of the squad's weapons pointed outward, the container complex resembled a battleship. It was a unique fighting position, to say the least. The platoon headquarters and the company command post took over a small compound within the base and immediately set about establishing communications with the platoon positions and the remainder of Brutal Nation back at Camp Marmal.

All Paratroopers had adequate cover and concealment and the buildings were in good enough shape to keep those resting out of the elements as the summer heat was now bearing down on that sunny day. We watched for the remainder of that sunny day as Chinooks brought in Afghan Commandos and their Special Forces counterparts. A UH-60 MEDEVAC bird landed and rested on the pad in which we had arrived. With security in place, we had provided freedom of movement for CJSOTF to build up combat power for their operations. A squad at a time would be the QRF element, with a skeleton crew manning security positions. CPT Blackburn and I agreed that in the event of the QRF being activated, he, myself, our FO SGT Demetrius Dasher, and our medic SPC "Doc" McCoy would accompany them.

That evening, we waited patiently for the signal. The SOF were out on mission and our QRF was staged adjacent to the CH-47 Chinook assigned to bring us in. SSG Lawson and I regularly made the rounds at the squads' positions. The men looked in good spirits. For starters, we weren't guarding a base within a base in the most liberal city in Afghanistan. Second, we weren't on the road. And lastly, we were supporting units that still conducted kinetic engagements. In essence, the war was still real, and the Kunduz operation made it feel like it.

For the next three days, we continued our security operations and observed the SOF teams lifting off and returning to base. Having submitted a packet for the Army Special Forces Assessment and Selection (SFAS), I longed to be a leader of one of these ODAs and I trained every day to prepare for the course. The letter of my acceptance had not arrived, but neither did 1LT Luke Ziller's letter, who had also applied. To build combat power for us, "Doc" McCoy was getting hands on training with the ODA 11D, a highly trained Medical Sergeant who utilized SPC McCoy to help treat the wounded Afghan Commandos.

Our communications with the TOC back at Camp Marmal were spotty at best, limited to a TACSAT radio that SGT Dasher operated. Talking with the battalion was even more problematic, confined to a Distributed Tactical Communications System (DTCS) radio which functioned more like a children's "walkie talkie." CPT Blackburn only received coverage at one position in our compound, a spot in a courtyard that he marked with a stick so that he knew where he could communicate. We limited any non-secured communications over our little local cell phones unless there was an emergency.

One afternoon, we were tasked to secure the nearby Kunduz Airport runway for incoming C-130's to bring in additional resources and enablers. A squad sized element was selected for the task. SSG Nooney's squad would have its two fire teams secure each end of the runway while myself, SGT Dasher, and "Doc" McCoy would position on the tower overlooking the entire airfield. It was simple enough, but we knew that one squad was woefully inadequate to defend an airport. Despite enemy threats in the area of operations, our task was more to keep curious locals off the airstrip and to allow birds to come and go without danger to people or aircraft.

A UH-60 Blackhawk set us down on the center of the tarmac. Kunduz Airport looked to be in great shape, despite being completely deserted. Within ten minutes, SSG Nooney's squad was in place on both sides of the runway and my small detachment had secured the top of the tower. After informing CPT Blackburn that we were in place, all we had to do was keep our eyes peeled and wait. It did not take long as C-130's started arriving. They landed, dropped their vehicles and personnel off, and took off. The inbound support for the CJSOTF mission quickly departed for the base without bothering us.

Late in the evening, we were informed no more birds would be arriving and to stand down. Our UH-60 transport was occupied supporting ARSOF missions in the area, so we moved back to our compound on foot. Throughout the darkness, under NVGs, we observed the desolate base along the mile long route to our destination. It was eerily quiet, and no signs of activity were witnessed. That walk back would be the only dismounted movement I conducted on that deployment outside of a NATO or Host Nation Government structure or compound. Regardless of how short or mundane it had been, i did it, on one foot.

An embarrassing moment occurred the next day over SSG Dreeson's position. A single stray round flew low over the compound, the entire base heard its whistle as it zoomed from the FARP, over the Brutal Nation command post, and towards SGT Tippit's position. Immediately, all Paratroopers went on high alert, thinking that somebody engaged SSG Dreeson's men on the tower. From what I later heard, the fuelers at the FARP were even scrambling for cover. No other round was heard for the remainder of the mission, but if you could have heard the chatter on the MBITR, you would have thought we were under a complex attack.

Everyone had been worked up over that one round, but thankfully cooler heads prevailed, and it was collectively decided to be a stray round from the city itself. I mean, the locals tend to fire AK-47s into the air for celebrations of all types, so it was entirely plausible.

Naturally, those without the Combat Infantry Badge wanted one, so SSG Dreeson's squad wrote up the sworn statements and the paperwork for the order. They were later denied however, as they did not meet the requirements for being engaged by the enemy. I would gladly take that one bullet over my head for the entire tour than a fire fight on any day. Not that I am a coward per se, I just don't want to deal with the consequences of combat unless I am 100% sure my team will win every time without casualties. And when that can't be guaranteed, and it never is, I want to ensure I have every tool and weapon available to hit them before they hit me. Don't go looking for a fight unless you're ready for the aftermath.

The next afternoon, we began to breakdown from our positions to return to Camp Marmal. Covering the FARP and CJSOTF elements as they departed, the Paratroopers from Brutal Nation were among the last to leave. It was a successful mission for us and proved beyond a doubt that the 82nd Airborne Division was mature enough and competent enough to support our brothers and sisters in the Special Operations community. I was proud of my team in First Platoon, who had acted honorably and without any hesitation to any tasks assigned to them. Without a doubt, this was one of the highlights of my tenure as a Platoon Leader.

For the rest of June, we continued our normal battle rhythm of SEC-FOR patrols and OCC-R security. Towards July, the OCC-R mission was officially shut down in Maser-E-Sharif and we departed the base, a welcome relief for us as the mission had been anything but exciting. Now our priorities consisted of SECFOR, training, and the inevitable redeployment to Fort Bragg, North Carolina which was scheduled for less than two months.

1LT Ausenbaugh spearheaded the redeployment efforts, coordinating to receive containers, customs inspections by the U.S. Coast Guard, packing equipment, and identifying theater provided equipment to turn in. Brutal Company did not have a replacement unit, so no relief in place was scheduled to be conducted. This should have been my first clue that the U.S. Government was winding down operations in Afghanistan.

The Fourth of July was celebrated on the base in modest fashion. Camp Marmal was a NATO base run mostly by the German Bundeswehr, so a big show was not to be expected. Regardless, CPT Blackburn and 1SG Goodart didn't want us to forget our roots so that evening, over awful pizza and non-alcoholic O'Doul's, the Platoon Leaders and Platoon Sergeants were given a screening of the classic film "A Bridge Too Far." While a far cry from the fireworks back home, it was a rememberable evening for me. I truly wished, however, to be back with Michelle for the festivities.

The real talk on base among our European friends, however, was the upcoming FIFA World Cup. Parties abounded on the base during a multinational contingent's home country's showing and cheers roared across the base when their players performed to their expectations. The U.S. team even had a good showing, with many of us joining in to watch the 90 minutes of fun on the Armed Forces Network.

While some of the contingents threw wild parties, we abstained ourselves and to my recollection, not a single Paratrooper violated the prohibition on drinking. Not that we needed the reminder though. In fact, when we first arrived on Camp Marmal, and discovered that many locations on the camp had "disco" and "fun" activities, 1SG Goodart was quick to forbid us from attending such events without authorization, which we never bothered to ask for since every single one of us knew the response.

I got to know my Platoon Sergeant, SSG Christian Lawson, extremely well during that summer. Many evenings were spent watching films in the MWR such as "Cool Hand Luke" or "The Outsiders." We confided how this deployment was different than anything previously experienced and when the MWR director announced that it was make your own ice cream night, Christian bemoaned, "Where are we?" He had seen a lot during his career and was involved in some heavy combat. Camp Marmal must have looked so alien to him when compared to the small outposts he operated out of with the 173rd Airborne and 82nd Airborne on previous deployments. I was lucky to have his experience and wisdom. Without a doubt, he made life easier for me.

Sometimes, we received news of events happening across the world. I would go into CPT Blackburn's office, and we would watch classified videos of terrorist attacks in the Middle East or drone feeds of real-world missions. Of consequence though, was a developing situation in Iraq and Syria. The Islamic State of Iraq and Syria (ISIS) had just taken the second largest city in Iraq, Mosul. An entire Iraqi Army Corps that U.S. forces had trained and equipped, had collapsed in the face of a few hundred ISIS fighters. The result was the fall of Mosul and the execution of tens of thousands of civilians and former Iraqi Government and Military Officials. We watched the executions and the attacks in silence. Little did I know that this event, happening two countries west of me, set conditions for a battle that would prove to be anything unlike I had ever faced just a few years later.

It was during this period that 1LT Ziller and I received news of our ARSOF applications. We were both turned down for SFAS due to a surplus of 18A Officers and those commissioned in the pipeline. Instead, we were offered to attend the Civil Affairs Selection Course which we both turned down. Michelle, as usual comforted me and gave me strength in the news.

While disheartened, I didn't fault the decision makers at the John F Kennedy Special Warfare Center and School. A friend of my friend, SSG Eric Myers, later informed me that, although I had a strong packet, the course couldn't accommodate an amputee at this time. That was a gut-punch to my self-esteem and ego but gave me the determination to prove them wrong by remaining in the uniform and going on to serve with the 82nd Airborne, leg or no leg.

By the middle of July, the SECFOR missions began to terminate as well. The engineers, about complete with their tasks, limited their missions to just a few Afghanistan Border Police stations and new projects in and around Maser E Sharif. One last mission for us was at hand prior to our departure. CJSOTF called CPT Blackburn again. There would be a second and larger operation out of Kunduz and this time, all of Brutal Nation at Camp Marmal was needed to help them achieve mission accomplishment.

The addition of Third Platoon in the planning process incurred some beneficial changes from the first mission. Now security was spread across the same compound at Kunduz, but Kamal's two squads now took over the adjacent area that formerly had been occupied by SGT Tippit. Additionally, QRF rotated per platoon per day so now no degradation of force protection occurred if the QRF was launched. The Company Headquarters element remained at the same location but was augmented with the FSO, 1LT Gabrielle Garcia and 1SG Goodart. 1LT Ausenbaugh and SSG Peveto held the fort down back at Camp Marmal, the XO being busy planning the re-deployment operation for the company.

SSG Lawson had the brilliant idea of bringing the platoon John Deere Gator UTV on the mission. Since we were riding in CH-47 Chinooks on this round, the Gator was modified to fit in the back of the aircraft and utilized for resupply and fire team movements around the compound or the adjacent airport. While dubious at first at the incorporation of the UTV, his idea proved brilliant in facilitating our movements. After attending the combined arms rehearsal for CJSOTF, CPT Blackburn briefed us on our mission specific details. Subsequently, I read the platoon in on our piece of the pie and went over some contingencies. Being that this was a carbon copy of the previous mission, not many questions were asked. The biggest addition to the plan was that the airport was to be secured daily during selected windows that the C-130s were flying into. The CJSOTF mission for Kunduz was to be even bigger in scale than the previous operation.

On the morning of the operation, the school bus sized helicopters carried Brutal Nation east towards Kunduz. The mood among the Paratroopers in the hanger was high that morning, we had less than a month prior to redeployment and this operation was a swell way to close out the deployment. We had suffered no casualties over the last seven months and were determined to keep it that way. SSG Lawson had the best seat in the house, riding on the Gator facing the tail ramp, giving him a glorious view of the Afghan countryside.

As two Chinooks touched down, we exfiled and quickly moved to our assigned positions. As Third Platoon made their way to their compound where the "battleship" position was, a chain and lock blocked the entrance at the gate. A quick set of bolt cutters quickly made short work of the obstacle before Kamal ordered his boys to clear the area. At the pace and tenacity of their movements, you would have thought Osama bin Laden himself was behind that gate. I can't fault Kamal for taking it as seriously as he did, but I would be lying if SSG Lawson and I didn't have a good laugh about it.

The First Platoon compound adjoined with the Company Command Post and to our surprise, one of the team even managed to get the air conditioning unit functioning in one of the buildings that served as a rest area. SSG Nooney and Dreeson later informed me that the Afghan Soldiers who occupied the base must have stripped whatever working components were left at their positions. SSG Nooney's building had lighting at one point, but all of the electrical components were now missing. Despite the degradation in comforts, Kunduz was still a heck of a lot better than being on the road.

The Gator proved its weight in gold during the airport security missions. While UH-60's brought in the main element, SSG Lawson or Nooney drove the mile to the airport's apron and met up with the team. From there fire teams were quickly shuttled to their opposing ends of the runway in record time. The hours were a lot longer this time around as C-130's landed and took off throughout the afternoon and well after darkness fell. A cool breeze fell over the airport, and with our assault packs, we relaxed as much as we could and took it all in. Paratroopers ate food from the PX or DFAC that they had brought, and everyone consumed the official energy drink of war, "Rip Its" or the theater hydration beverages called "Mega Sports."

A nameless operator met up with the headquarters element in the air traffic control tower and started shooting the shit. He then advised us to put down our NVGs and stare off into the darkness in a particular direction. AC-130 Specter Gunships were supporting the CJSOTF mission and the tell-tale sign of their 105mm cannons or mini guns ruining some insurgent's evening broadcasted to us a magnificent light show. Lasers from weapons systems far off into the night were also seen and the muffled sounds of explosions in the distance were heard. While I never received his name, I thought it was cool of that CJSOTF bro to come see the team supporting his outfit and orient us to the missions that we were not read in on.

Kunduz for the next few days went on much of the same way, guarding the arrival of logistics from the airport, pulling security on the compound, or being ready to respond on our nights with a QRF element. Still cognizant of the insider threat, we moved in buddy teams outside of compounds when visiting other positions. At one point, when 1LT Garcia was making his way to visit another location, the command element asked him who he was taking. Our FSO, a religious man who also catered to the spiritual needs of Brutal Nation, held two MBITR radios in his hand and responded casually with, "Don't worry, sir. I've got Jesus and two AH-64 Apache's," before swaggering off. I guess one couldn't argue with that logic.

A lot of down time was to be had on those hot days at Kunduz. "Doc" McCoy was often seen giving training to the Paratroopers, including administering IVs to each other to force hydration or even combat life-saving skills under NVGs. It was also this time at the command post that my left leg was giving off a very bad smell. We had not showered for days and only baby wipes were available to wipe off the moon dust that permeated everything in Northern Afghanistan. 1SG Goodart wasted no time telling me that my leg smelled like death every time I took off my prosthetic to let the skin breath. It got so bad that other Paratroopers stopped resting in any room I was in. I resorted to washing it with potable water and drying it. It wasn't until much later that I realized I could have just rubbed a deodorant stick on my residual limb to control the funk. At least everyone had a good laugh about it.

The Gator earned meritorious praise by the CJSOTF medical team for its work in facilitating MEDEVACs of wounded Afghan Commandos. "Doc McCoy" and SSG Lawson often drove straight up to the UH-60 MEDEVAC Blackhawk and transferred litters onto the back of the Gator for immediate transport to the Special Operations Surgical Team (SOS-T) that was on site. Multiple Afghan Soldiers were swiftly transported while "Doc" gained more experience. It was a win-win and bolstered the respect that Brutal Nation garnered.

The downtime also impacted morale when thoughts of home popped up. We had been gone for months, and being back home amongst family and friends was the most sought-after fruit of our homecoming. The Kunduz trips were the longest periods on that deployment where I did not talk to Michelle.

In 2012, you had to physically go to the MWR tent, sign in for a booth, go to the telephone, type a code, and then when prompted, type the number on a phone card with pre-bought minutes on it. There were no face-to-face meetings via teleconference.

Two years later, Facetime and Skype were available that eliminated the need to purchase phone cards. The only caveat was you must buy a Wi-Fi puck for your hootch and then buy the data cards to continue the Wi-Fi usage. While spotty at times, the data cards were prevalent everywhere, so all Paratroopers had the ability to get on the internet from their own private quarters.

Michelle had done incredibly well in my absence. Her second deployment experience, while much longer, at least had the possibility of me coming home in one piece. Even more important though, was that we had both agreed to extend our family from the two of us and our puppy, Aussie, to trying to conceive. At Walter Reed, the extent of my injuries meant that having children would be difficult, if not impossible. For the time being, I could only focus on the mission and patiently counted down the days until homecoming.

Others in Brutal Nation had not been as lucky. Our company armorer received the news during the second Kunduz mission that his wife was leaving him. Everyone consoled him as best they could, the biggest nightmare of all married Service Members had occurred among our own and for many, either reassured themselves of the stability of their own situations, or for others, increased their own anxiety about the insecurities they felt between themselves and their loved ones.

The last day of the Kunduz mission entailed the complete breakdown of all locations utilized for the operation and movement by air to Camp Marmal or the supported unit's respective base. Brutal Nation was among the last to leave, guarding the CJSOTF assets as they departed for the airfield or lifted off straight from the compound's landing pads. Breakdown for us took no time at all as we were Paratroopers, expeditionary by nature.

Before leaving for the airfield, one of my Paratroopers, PFC Thrasher, presented a large metal key that he found in one of the old German machine shops on the base. It was a ridiculously large skeleton key that was handed over to 1SG Goodart in a small ceremony with SSG Lawson and Thrasher. "1SG, the Key to Kunduz is yours," the young Paratrooper proclaimed to the bewildered 1SG who graciously accepted the key, fully understanding the irony of the symbolic gesture.

As dusk fell, our chalks moved into pickup zone (PZ) posture to be picked up by the birds to return to Camp Marmal. Most of First and Third Platoon was slotted with UH-60s while the remainder, to include myself, rode back on a CH-47.

As my guys started walking to the bird, rotors spinning and all, I noticed my RTO, PFC McCarrick still had his long antenna up on his manpack radio. Quickly seeing the hazard, I ran over to him and had him collapsing the antenna, thus avoiding any possible damage to McCarrick, the helo, or the radio. Looking back at it, I should have noticed sooner when everyone was in line for PZ posture.

As the sun finally slid under the horizon, the giant bird with its distinctive dual rotor sound, lifted off. I quickly fell asleep in a seat forward of the fuselage while SSG Lawson enjoyed the view on the return in his Gator. Despite the whine and whooping of the turbines, I had no trouble dozing as the aircraft was pitch black besides the blackout lights.

My last mission as a Platoon Leader was coming to an end. I was destined to take over from 1LT Ausenbaugh as the Brutal Nation Executive Officer upon our redeployment to Fort Bragg. He would move on to the Battalion S4 position, managing the organization's logistics. Although Kunduz would soon become memory, the experience gained followed me for the rest of my career.

Brutal Nation was getting ready to return home. With the loss of patrols on our plate, the Paratroopers routine for the last couple of weeks comprised of working out, eating, a daily huddle for the leaders, and any tasks that the XO needed us to do to facilitate our homecoming. Containers were loaded with organic property, inspected, and sealed. Customs agents checked all our personal equipment and property to ensure no contraband was present before signing off on the Paratroopers.

Our lumbering MRAPs, which had faithfully carried us throughout the northern provinces, were destined to a grim fate. Unable to be shipped back to the United States due to costs, they were turned in for scrapping. Misc. components were ripped out and packaged into kicker boxes, sensitive items were collected, and vehicles given a final wash before being turned over to contractors for their destruction. It was a sad and inglorious fate to those vehicles that must have cost American taxpayers hundreds of thousands of dollars each.

Our flight out of country was scheduled for August 6. First Platoon was Main Body One which meant first out the door after the ADVON force departed about a week prior. After flying out of Afghanistan, the strategic lift aircraft would touch down in Romania where we would await the commercial contract flight back to Fort Bragg. If Transportation Command and luck worked together, I would be home by Michelle's birthday. By now, my bags were packed, and the waiting game begin. To pass time, I either worked out at the gym or hung out at the platoon tent or company CP.

On one of the last evenings prior to departure, 1LT Ausenbaugh was working on a packet for a course in lieu of the Maneuver Captain's Career Course that all Infantry Officers attended prior to their company command assignments. The course was the United States Marine Corps Expeditionary Warfare School in Quantico, Virginia, a year-long academic course focused on educating Marine Officers on amphibius and joint operations with the Marine Air Ground Task Force or MAGTF.

Incredibly curious, I immediately built up a packet and solicited letters of endorsement from CPT Blackburn, General David Rodriguez, and my State Senator Mitch McConnell. After fulfilling the initial packet requirements, I contented myself to the wait and see game as I still had an entire year as a Company Executive Officer to look forward to.

On a sunny morning in early August, the Paratroopers from First Platoon, Brutal Nation of the 2nd Battalion 508th Parachute Infantry Regiment walked over to the PAX Terminal for the flight out. 1SG Goodart accompanied us while CPT Blackburn and Kamal gave us our farewells prior to heading out. As we boarded the giant maw of a C-17 Globemaster, I sat next to SSG Lawson. My mission complete in Afghanistan, I finally breathed a sigh of relief. I had done what I came here to accomplish.

A short stop in Kandahar Airfield to pick up Delta Company was our only detour. I didn't leave the plane but looked out as the tail ramp opened. This was the first time I had returned to Kandahar since I lost my leg. My former outfit, Delta Company, loaded aboard. Just like us, not one of them had been wounded on the mission and everyone was coming home intact. The Lord had truly blessed Task Force 2-Fury.

As we taxied for takeoff, Paratroopers hailed their friends whom they hadn't seen in over eight months. Packed like sardines, we couldn't really get up but contented ourselves to shouting over the roar of the engines as we took off into the sky.

The flight to the Mikhail Kogalniceanu Airport did not take long. We touched down in darkness and were immediately hustled to buses on the tarmac for transport to the base that the United States was leasing from the Romanian Government since the Manas Airbase at Kyrgyzstan had closed for business earlier in the year. Trees, grass, and bodies of clean water, the first we had seen in many months, greeted us through the dimly lit road taking us to the living space area where we would reside for the next 24 hours.

The next day we were free to roam the confines of the base. I bought a homemade chess set from a gift shop and played with one of the Captains in Delta Company, CPT Sam Hammer, a graduate of Princeton and future Harvard Law School graduate. Needless to say, I do not think I beat him.

On the morning of August 8, we boarded a Boeing 777 flight contracted by Atlas Airlines. The trip was nonstop, no refueling required, which meant I would get back on that late afternoon to wish my young wife a happy birthday. There was no first class. We sat like sardines for the long flight, but few complained, our homecoming was soon at hand.

Several hours later, we touched down at Pope Army Air Base adjacent to Fort Bragg. Walking off the plane, we were quickly put in formation with berets on and guidons out front. A detail of Paratroopers from our rear detachment began unloading our bags from the aircraft and we waited for what seemed like eternity. Once the conditions were met, we finally marched in formation to the PAX shed at Green Ramp where tens of thousands of Paratroopers had marched into before us upon re-deploying from their conflicts.

The All-American Division Band was playing the usual triumphant melodies as we entered the building to the enormous roar of cheering families and friends. "Mark Time March," "Group Halt," and "Right Face" were ordered, the Paratroopers instinctively responding to the commands.

As I turned right and assumed the position of attention, I saw her. Michelle was beaming at me and wearing a beautiful sun dress. She looked stunning and was completely dolled up for the occasion. I was indeed one lucky man.

As the Division Commander, MG Nicholson gave us some kind words of congratulations and welcome home, I almost lost my patience. I couldn't stand seeing her any longer and not embracing her. Then, the moment came when we were told "dismissed." Michelle broke through the barrier and jumped on me. I embraced her and kissed her better than I had done when we were before the alter. I was completely ecstatic, no more lonely nights or days away from my wife. We overcame so much over the past two years and this moment was proof that anything could be overcome.

After our initial reunion, all returning Paratroopers were shuttled to their battalion footprints. Since the 4th Brigade Combat Team had been disbanded, we were now in the 2nd Brigade's area of operations and thus, new Company Operating Facilities (COF).

After turning in our platoon equipment and weapons, families and returnees met behind the COFs for the final announcement that the sensitive items were secured. By now, it was nightfall, and we were growing ever more impatient to leave. When we were finally released, my old friend from EKU, 1LT Caleb Wood helped me load my two tough boxes and duffle bags into my truck. Starving, we went to our favorite Mexican Restaurant, El Cazador in Fayetteville for an epic round of food and my first alcoholic beverages in almost a year. I didn't drink much though, I needed to be ready for the evening. I do not think I have ever made such passionate love in my entire life. Happy birthday, Michelle.

Josh reunites with Michelle at Fort Bragg after his tour in Afghanistan, (AUG 2014)

CHAPTER 17
JUMPMASTER
SEPTEMBER TO OCTOBER 2015
FORT BRAGG, NC

The transition back home for me was easy due to the efforts of Michelle to ensure our home was ready for my arrival. The workload in Brutal Nation that fall was minimal due to the fact much of our equipment was in the middle of the Atlantic Ocean enroute to the United States. Thus, the first couple weeks after returning home comprised of half-day schedules and four-day passes to give Paratroopers time back with their families.

In October, we would receive a two-week block of leave before restarting the process of preparing for war. A rotation to the Joint Readiness Training Center (JRTC) at Fort Polk, Louisiana was scheduled in late April of the following year, so the unit had a finite amount of time to reconstitute itself and train from individual to platoon echelon. In essence, enjoy your fall now because winter and spring is going to punch you in the face.

Not more than two weeks after returning home, I was on another flight overseas, this time with Michelle. During the deployment, I had been interviewed by a British journalist working with NATO who was doing a story on individual Soldiers across the participating countries.

For two weeks, she imbedded with First Platoon and even accompanied us on missions. After she had left, I later received an invite to the 2014 NATO Summit in Whales, Great Britain for a ceremony about the piece she did called, "Return to Hope: NATO's Journey in Afghanistan," which in addition to me involved five other Service Members from different NATO countries.

I was alone in this endeavor, so I requested that my wife be allowed to accommodate me, which they accepted and generously funded. As a representative of not just the 82nd Airborne Division and the U.S. Army, but my entire country, I was briefed prior to departing by division public affairs on what to do and say in certain situations. I did not want to embarrass my country, so I was all too happy to receive some coaching.

Arriving in Bristol, England, my wife and I took a bus and then train to Cardiff, the capital and largest city in Wales. This was the first time Michelle had flown over the Atlantic and probably the third time she had ever left the country apart from some missionary work in Mexico and our honeymoon to Aruba. She was wild eyed the entire time, despite the jet lag from the long flight.

The scenery once we left Bristol was very picturesque and aspects of it even reminded me of home. We had no issues navigating the mile distance to our hotel in Cardiff. As usual, Michelle overpacked so carrying the excessive baggage was the most of our issues.

Our trip was only about four days long, so resting that first afternoon was not an option. As soon as we were settled at the hotel, we were immediately whisked to the location of the NATO Summit. My interviewer from Afghanistan was there to greet me and walk me through the credentialling process. For the day of the actual summit, Michelle would remain at the hotel due to clearance issues. As she was absolutely exhausted from the flight, a whole day of sleeping in and ordering room service did not sound too bad for her. On the last day, we planned on sightseeing in Cardiff to make the best of the trip.

I could never confirm, but there was a rumor that a pair of generals from the Afghanistan National Army who had been invited to the Summit caused a controversy upon their arrival. Immediately after landing in Great Britain, they claimed asylum to avoid returning to their home country. How ironic I thought, to be representing your country during one of the biggest events for NATO, only to defect once the opportunity presented itself.

This should have been a huge indicator to me that the transition in Afghanistan from NATO led operations to the home government was not going as well as I thought. Despite the scuttlebutt of the general's act, there were no other indicators that things were array, and the summit went off as expected.

After the showcasing of the "Return to Hope" piece in a ceremony, I was free to my own devices. I did not plan on staying as my wife was back in a hotel by herself, but I did have to stick around long enough to be cordial. An open bar was at the manor where the ceremony was held. I was in my Army Service Uniform, so I stuck out like a sore thumb in a sea of men and women from the Queen's Armed Forces.

I decided to do the one thing Soldiers can agree on. I bought a Royal Officer a round of drinks and we shot the shit on our lives and service. Every time an enlisted Soldier walked by, I was greeted with a "Sir" and salute before they departed. It was unique examining the dichotomy of the enlisted and officers of the Armed Forces of Great Britian. There was a unique separation and courtesy that must have extended back hundreds of years throughout their history. I tried to buy one of them a drink but was politely informed that he would have been in some trouble should he accept.

After a couple of hours, I returned to pick up Michelle for the evening dinner. She did not look amused at having to be left in the room but luckily it was almost over. We dined with the other members of the "Return to Hope" team. I was plied with endless rounds of wine and Italian food, all exquisite and delicious. After dinner, I tried to get some sleep but was still overcome with jet lag. She had no issues however, and to this day, I envy how easily Michelle can fall asleep while insomnia continues to plague me to the present.

The day prior to departure was our own. We contented ourselves to exploring the Welsh capital and its environment. A history major, Castle Cardiff was a must for me, and the tour was superb. I highly recommend a visit to Whales, Great Britain for any adventurist as the sights, history, and people were simply amazing, not to mention the cuisine. The flight back to the States, my third overseas flight in a month, marked an end to my international adventures for quite some time.

At this point, a month since I had returned home from war, I was becoming restless. I had skipped the September class of Jumpmaster School to attend the NATO Summit. Prior to assuming my duties as the XO, I would attend this course in the hopes of becoming a bona fide Jumpmaster in the U.S. Army. But prior to that, something bigger happened.

Prior to attending the 82nd Airborne Division's Advanced Airborne School or commonly known as Jumpmaster School at Fort Bragg, all Paratroopers first had to have twelve successful exits from a high-performance aircraft (i.e. C-130 or C-17), a minimum rank of E-5, and must be on jump status for a specific amount of time. In addition, they had to pass a one-day entrance course called White Slip where Paratroopers had to pass two tests such as a 25-question nomenclature exam on specific pieces of the parachute and rigging a ruck sack with their air items in a specific amount of time. Those that passed both events were given a signed piece of paper or the "white slip" authorizing attendance at Jumpmaster School.

On the day I attended and passed White Slip, I received a phone call from Michelle. Michelle rarely called me while I was on duty as I was often too busy or in the middle of an area of no service to respond, so I knew it was important. She sounded both frantic and nervous when I picked up. "I'm pregnant!"

So, despite the odds, we were going to have a baby. And to top it off, I had only been home for a month. That didn't take any time at all. I was so excited that it was hard to focus on the two exams at White Slip, but nonetheless persevered for the coveted piece of paper. Despite what doctors said, I was going to be a daddy next spring. My focus now wasn't just on being a leader of Paratroopers and a husband, but to prepare for one of the hardest and most rewarding jobs of all, fatherhood.

I started Jumpmaster School in October. Three weeks long, I attended when the rest of Task Force 2-Fury was on block leave. Luckily for me, I would take leave upon graduation. Jumpmaster School is by no means a pushover. Traditionally, over 50 percent of the class fails, mainly due to the Jumpmaster Parachute Inspection (JMPI) and recycles to a later class.

However, it is a gentleman's course. There are no excessively early first calls, students are typically released at the end of the duty day, and cadre never yell or "smoke" students as one might see in Ranger School. In fact, the cadre were about as easy-going and professional as I had ever seen. The reason for the demeanor was to encourage learning and to allow the rigors of the exams to be the point of stress, not the instructors themselves who worked hard to transfer their knowledge onto the students.

The first week consisted of a nomenclature exam, learning the sequence of inspecting the T-10 and 11 series parachutes with and without combat equipment, and a written examination. The second week continued to build upon the inspection process for JMPI, a safety duties examination, and reciting all the pre-jump script.

On all exams, students needed 70 percent to pass and 80 percent or higher allowed for an additional attempt at JMPI. That means by scoring an 80 percent on the nomenclature and written exam, you could earn two additional attempts to pass JMPI or be "reentry qualified."

The final week was dedicated solely to JMPI. The exam allowed for three attempts to pass, and students attempted once per day allowing for failures to go back and study what they missed. The exam was as follows, you had three jumpers (two without combat equipment (Hollywood) and one rigged for a combat jump. You had to inspect all three jumpers within five minutes while adhering to the sequence, time, and finding deficiencies. Miss one too many deficiencies, you're out. Break sequence, you fail. Bust time, goodbye.

It was hard but we were given hours and hours of dedicated instruction and practice to figure it out. Reentry qualified students had one or two more attempts, so everyone wanted those high-test scores to reduce the stress.

The worst part of the course was the most beneficial. The entire class stood in a giant circle. There was a T-10 or T-11 parachute, rigged ruck sack, Modular Airborne Weapons Case (MAWC) with a rubber duck M4, and a T-10 or T-11 reserve parachute. Luckily for us, the actual JMPI exam was just the T-11 series parachute since the T-10 was being phased out. Half of the class would don the parachute while the other half inspected. Prior to anything we did or touched, the instructors gave us a thorough block of instruction. After rigging and inspecting the "Hollywood" jumpers, we switched out and donned the parachutes ourselves.

Being an amputee did not help me during the block of instruction where, for what seemed like an eternity, I had to squat and hold up a rucksack while the instructor taught us how to inspect the jumper at this point in the sequence. My leg cut deep where the carbon fiber met skin. It was excruciating and took every ounce of willpower to not lower the ruck and raise.

Sweat poured from my brow, and I was soon saturated. My left limb was on fire, and I knew I could only keep this up for so long. Finally, the instructor ordered "recover," which we all gratefully did. That point was the single hardest moment for me in Jumpmaster and I passed that reckoning. Now I just needed to pass the course, so I wouldn't have to endure that hell ever again.

When off duty, I practiced for the JMPI exam on Michelle by shadowboxing the sequence. Only two months pregnant, Michelle wasn't showing enough to "mimic" the placement of the T-11 Reserve Parachute, but she was a team player all the same. Apart from this, my free time in the evenings was spent on rehearsing the pre-jump script, a pages long speech of the "Five Points of Performance" and actions to take in various contingencies such as ground wires, tree landings, and water landings.

The written exam was studied for around an hour each evening from the over one-hundred flash cards I developed. I was feeling confident in all the aspects of Jumpmaster School that I could master. The written exam and pre-jump had no secrets. Either you knew the data or didn't. JMPI was a whole different monster. The only constants were the two "Hollywood" and one combat equipment jumper and the five-minute time hack. The variables were the dozens of potential minor and major deficiencies the instructors had learned how to cleverly hide so that a cursory inspection would lead to heartbreak.

For hours, each day of the first and second week on those warm fall days, we practiced the JMPI sequence under that pavilion. Once we mastered the sequence on the "Hollywood" and combat jumpers, the practice sessions began to introduce deficiencies into the lineup. Every jumper had different deficiencies. For an hour, half the class inspected jumpers rigged in harnesses. For the other half, you were being inspected. It is no fun being the person in the chute, having many sweaty hands and sometimes bodies lacking in anti-perspirant rubbing up against you. Some got the hang of it fast. Others needed as much practice as they could get.

During the second week, the written exams and the pre-jump exams were given. Acing the written exam in a classroom setting, I was still reentry qualified and had up to five attempts on the JMPI exam the following week.

On a cool early morning, all the students were formed up by their chalks and a designated instructor took them to a specific area at the

Advanced Airborne School. A student was called away out of earshot from the other students and was briefed to pitch pre-jump. We could see the various performances given. Some students stayed in one spot. Others looked like they were having a heart attack when they stumbled or forgot a section. Some walked in circles as they delivered their script.

When my turn came, I walked over to my instructor, a tattooed Master Sergeant named McAlvin. As with the instructors on the Fort Benning Basic Airborne Course, he wore the black hat with the unit crest, his jump wings, and rank. After briefing me on the exam's task, conditions, and standard, he asked if I was ready. After confirming, he told me to begin and started his twenty-minute countdown.

The entire script came naturally. My father had done the exact same school decades ago and was a Master Jumpmaster. Every aspect was briefed by the book, with no exceptions. Time seemed to stand still for me as I briefed MSG McAlvin on pre-jump. When I completed, he stopped the clock. I passed my final gate and scored well over the minimal points required to be re-entry qualified.

I had five attempts at JMPI now, so the stress level during the following week was minimalized. Apart from the safety duties exam which posed no challenge, the rest of the week was dedicated to JMPI. After mastering the sequence and being able to identify deficiencies with ease, the practice circle was abolished, and we began to mimic exam conditions on three poor Paratroopers who were volunteered to act as our guinea pigs.

The introduction of the time hack threw a lot of students off, including myself. It almost seemed impossible to get through five jumpers. But slow is smooth and smooth is fast. After hours of rehearsing under testing conditions, we were shaving off time and getting the hang of it. The following week was the final test for our class.

That first Monday of the third week was spent as a refresher after the weekend break doing countless JMPI circles and mock exams. On that afternoon, we were given a final brief on the expectations and layout for the exam the following day.

Usually, two Paratroopers I had served with in Delta Company, 2-Fury back in 2012, SSG Evangeliste and SSG Maravi, and I would take our lunch breaks at a joint called McKellar's Lodge on base. A buffet style restaurant, on one day we ate our weight in North Carolina barbeque prior to returning to the schoolhouse. The lethargy we obtained that afternoon killed any motivation in the JMPI circle, so I abstained from dining out again unless it was a light duty day. Lesson learned.

Those that passed their exam would conduct landscaping duties around the schoolhouse such as cutting grass or pulling weeds. Those that never cut the grass were those weeded out and were cut. So, an informal motto, apart from the class motto "Kid Tested, Mother Approved," was overheard throughout the third week. "Cut grass or get cut."

As everyone departed the Advanced Airborne School parking lot that afternoon, a sense of anxiety and foreboding was in the air. For some students like me, this was our first attempt and thus, had no clue what to expect. For others, this was their second, third, or fourth attempt. One female Captain had been in the course over five times.

Tuesday was the first attempt at JMPI. Wednesday was your second try if you failed to pass the first, which almost everybody would. Thursday was attempt number three. And Friday was attempts four and five for the reentry qualified students.

About five stations were set up around the compound. To keep the exams fair, all stations were rigged with the exact same deficiencies for each iteration. The deficiencies were switched out for each subsequent exam so there were no freebies. The entire class was herded into the classroom where they sat. While we were free to talk, not many did, choosing to focus or meditate on the task ahead. To keep students who had recently tested from sharing information, cell phones were forbidden on the grounds under penalty of disenrollment from the course.

Nobody knew the order of who was to be tested first. It was all random and as completely fair as possible.

Soon, the door leading into the classroom opened and a black hat announced the start of the JMPI exam. Five roster numbers were called, signaling the identified students to standup and follow him out. Then as suddenly as it opened, the door closed. For some time, this was our only contact with the outside world. The wait was excruciatingly long. Apart from the five minutes of the test, every station of three Paratroopers had to be rehabilitated and dressed up prior to the next tester. Finally, my roster number was called and four others and I followed the black hat out. Once outside, he repeated a roster number and motioned the direction of the student to take towards his station. My first station was at one of the giant pavilions with the rubber flooring.

Cautiously, as if being led to the execution site, I walked up to the exam station where four Paratroopers waited, three jumpers and one grader. I was given the usual brief of the exam's task, conditions, and standards. He advised me not to break sequence, bust time, miss deficiencies, or make up deficiencies that didn't exist, or "fishing" as the instructors called it. Understanding the task, I announced that I was ready and turned around, facing away from the first jumper to be inspected. "Begin!"

I took a deep breath and turned around to face my first jumper. Starting with his Advanced Combat Helmet, I began the inspection sequence. Tracing the inspection points with hands, I ensured my eyes always met the piece of equipment where my hands lay. Completing the helmet, I moved to the canopy release assemblies, initiating the inspection with a hard knock on the steel component to ensure it was properly secured. I grabbed the main lift web assembly on the parachute harness and put my face as close as possible to the side of the canopy release assembly, no debris present. I then focused on the main lift web, one side at a time, tracing the equipment with my hand and inspecting the tuck tab assemblies that help the Paratrooper size into the parachute. Both tuck tabs were properly fitted and assembled. So far so good.

I then inspected the jumper's chest strap, waist band, and T-11 Reserve Parachute. As I found deficiencies in the equipment, I called them. After finishing the reserve, I raised his parachute and told the jumper "Hold squat," the signal for him to grab his reserve and squat for the leg strap inspection. Once completing the inspection sequence on both leg straps and ensuring the ejector snaps were properly fastened, I issued the command of "recover," allowing the jumper to lower his reserve and stand back up.

Finally, I inspected the universal static line modified, the yellow rope like piece of equipment that is attached to the aircraft's static line and activates the main parachute when the jumper exits the aircraft. I told the jumper to turn and proceeded to inspect the remainder of his parachute by tracing his static line to the main curved pin. Upon completion, I checked the riser assemblies, pack tray, and the diagonal and horizontal backstraps. I finished the first jumper by tracing his saddle and gave him a slap on the backside and called for him to "recover."

All of that for just one "Hollywood" jumper and as of writing this, I skipped about 90% of the details that are in the script for the course of instruction. It's a lot to remember and rehearse but it just may save someone's life. Time was going and I still had two more jumpers, to include the combat equipment jumper which entailed a lot more to the inspection sequence due to the addition of the rucksack and MAWC.

I wanted to move faster, but figured I had four more attempts at JMPI. I just wanted to ensure I had this down at a smooth rate and didn't miss the deficiencies rigged throughout the exam, like landmines in Afghanistan.

The second jumper was the combat equipment jumper. I started the inspection sequence just as before. I called deficiencies as I found them and kept up the inspection. The MAWC containing the M4 Carbine was quick to inspect. I ensured it was properly secured to the parachute harness and that the weapon was properly secured. The inspection of the air items on the ruck was even easier. Finding twists or misrouted webbing, I called out the deficiencies and moved on. I could tell I was behind as it seemed an eternity had elapsed for the first two jumpers. The instructor was silent, giving time checks every minute.

The final jumper was "Hollywood" and the inspection went by as proscribed. As I smacked his rear end and gave him the "recover" order, the black hat called time.

Immediately he went into his feedback. I had blown past the allotted five minutes for the exam and had missed one major deficiency (both of which automatically failed me). The major deficiency was an offset protection cap which to me seemed invisible to the untrained eye. This deficiency failed most of the class that day I would later find out. Regardless of the setback, I confided in myself that I did not fail in the sequence and that I had purposely performed slower to ensure I didn't miss anything. Tomorrow was another day, and I was still in the running to obtain the coveted title of Jumpmaster.

The next day was a clone of Tuesday in that the JMPI exam was in full swing. That morning, we did some practice JMPI circles to remedy our performances yesterday. Some students had to work on sequence. Others had to increase their pace to make the five-minute time hack. Deficiencies plagued the remainder. Some students had to improve on all three types of exam obstacles. Only about three of the class of almost sixty had passed so the JMPI circle was no different.

That afternoon, the second exam attempt was initiated. Mirroring yesterday's execution, the roster numbers were randomly called. This time my wait was shorter. I had my sequence down and my pace was fine. I knew what I needed to do. It was now or never. I rose quickly as my roster number was called. I was motioned to three chairs in front of the Advanced Airborne School Headquarters where a black hat was waiting. After the exam instructions brief, I was again asked if I understood and was ready. I turned around and waited for the command. "Go."

I turned around and began my second attempt at JMPI. I moved faster than the day prior but was smooth in my inspection. I felt great as I sailed through the first jumper and immediately began inspecting the combat equipped Paratrooper. I was finding deficiencies left and right, as if someone had placed a beacon strobe on each piece of equipment that was improperly rigged. Every thought in my mind vanished as I focused on the inspection process, as if my father's legacy or my unborn son's life depended on it. Moving on to the last jumper, I could sense pairs of eyes on me. Hushed voices were heard as I traced the static line. Was I missing something? Did I miss an important step in the sequence? And that's where I saw it. I had missed the static line being mis-routed through the outer static line stow bar. Still keeping my hands on the static line, I called out the deficiency that I had narrowly missed only seconds before. I did not violate the sequence because my hands never moved back to the deficiency, and I continued inspecting. I believe it was the closest call in the history of Jumpmaster School.

As I smacked the final Paratrooper's ass and announced "Recover," the time was stopped. My audience was none other than 1SG Woodruff, my former 1SG at the Walter Reed Warrior Transition Unit (WTU). He was silent as the instructor on my opposite side briefed me on my performance. My sequence was fine. I found all minor and major deficiencies, missing nothing that would have failed me. Then, the black hat started telling me about my time and how there was only five minutes allotted for the exam. My heart sank. He showed no emotion as he explained how I should be competent enough to pass this after over two weeks of training. There was no excuse to bust time. He then tried to comfort me in the fact that I had found all the deficiencies and that my sequence was perfect. I had three more attempts and he assured me I would get it the next time.

That's when he held up his stopwatch and grinned. 04:40 was on his stopwatch. He had been messing with me for the last couple minutes for his amusement. I couldn't believe it. "So did I pass?" I asked in disbelief.

"Yes, Jumpmaster." I exploded. I couldn't believe it. I must have hugged the hell out of that black hat and lost all military bearing.

1SG Woodruff was also grinning his ass off, the prank surely being at his insistence. "Congrats, Jumpmaster!" I still couldn't believe it as I floated to my truck to let Michelle know the news. I was still shaking on the way back and must have called everyone to let them know I passed JMPI. I called CPT Blackburn. I called my father. I called SSG Lawson. The single hardest exam I have ever executed in my life was over.

The last two days of the duty week for me consisted of doing landscaping around the Advanced Airborne School compound while other students retested for their subsequent iterations. The stress I had leading up to the exam now absent, I now looked forward to the only obstacle prior to graduation. One aspect of Jumpmaster School is the Practical Work in the Aircraft (PWAC) exam where students execute the actions of a Primary or Assistant Jumpmaster while in flight. Consisting of hand and arm signals and the paratroop door safety checks, it is among the easier of the exams due to the fact all the students had seen Jumpmasters execute this ritual many times on airborne operations.

Upon completing the exam, the student executed a high-performance exit from the aircraft and descended back to Earth under parachute canopy. PWAC is typically done after JMPI due to the number of students having been weeded out, but sometimes it is prior or much later due to availability of aircraft. For Class 01-15, PWAC was scheduled the subsequent week just prior to graduation so there would be no extended purgatory between JMPI and graduation.

A humorous anecdote took place the previous weekend that bears mentioning. The Army Ten Miler Race was conducted the Sunday prior to JMPI exam week. Previously, my old friend, Harvey Naranjo, COTA, from Walter Reed and leading facilitator of adaptive sports and recreational therapy for the patients, had signed me up for the race which I had planned on running. However, not wanting to chance failing JMPI for lack of studying that weekend, I opted to stay at Fort Bragg and allow one of the medical staff to take my place. Here's the interesting part. That staff member had two legs and was a very good runner. Somewhere along the line, my category of single below knee amputation was not revised to reflect the new runner, or the name.

After the race concluded in Washington D.C., I started receiving random messages and voicemails of congratulations. MSG Joe Kapacziewski, a famous Ranger who had also lost his leg in combat sent me a long message of congrats and that's when I realized what had happened. I called Harvey and we had a good laugh about it when he told me the runner who took my packet ending up winning in my category.

What I did not realize was that the U.S. Army Public Affairs Team, 82nd Airborne Public Affairs Office, and the local news was getting ahold of the story. What made this turn sour, however, was that the Battalion Headquarters was getting wind of it and asking CPT Blackburn about it. Since I had not submitted a mileage pass for DC, leadership had the perception I violated the mileage policy to go run in the race.

It took some conversations and explaining how I did not run the race and had remained at Fort Bragg to focus on Jumpmaster School. I believe the chain of command was dubious of this explanation but the fact that I passed JMPI quashed any further digging into the matter. Had I failed JMPI, there would be no doubt I would have to involve Harvey in explaining that I never left the 250-mile radius of Fort Bragg.

For some time after that, my peers and some of the humorous leaders in the organization gave me hell for having someone else run for me so I could take credit for the win. Looking back at it now, it does seem sketchy at best on how this all played out. Regardless though, I'm setting the record straight by telling you that I did not race, let alone win in any category in the 2014 Army Ten Miler.

By the time the last of the reentry qualified students passed or failed their fifth and final attempt at JMPI, the class size had shrunk by roughly fifty percent since the start of the course almost three weeks prior. After cutting grass, weed eating, raking pine needles, and cleaning surfaces for a couple of days, the remainder of the class was gathered to go over the final days of Jumpmaster. All that remained was PWAC and graduation next week. We were to report for the Initial Manifest Call at the schoolhouse on Monday to practice for the exam which was scheduled for Tuesday. Graduation would be on Thursday after a day of cleaning and out-processing. After a quick round of patting ourselves on the back for passing JMPI and thus, remaining in the course, we were released for the weekend.

PWAC entailed another joke by the black hats on me. I don't know why I draw so much attention to myself (must be the leg) but I was the target of many "pranks" if you would. One of the rules of PWAC was that once you stood up and moved in the vicinity of the paratroop door to conduct the actions inside of the aircraft, you were not permitted to move your feet. Moving your feet under any circumstances warranted an automatic failure on the PWAC exam and thus, dropped from the course.

As I sat in the aircraft rigged up in the T-11 Parachute, I had the opportunity to watch my peers test in front of me. I had about six or so jumpers ahead of me performing the duties of the Primary or Assistant Jumpmaster, so it was good to see the exam unfold multiple times prior to my execution. As the jumper completed his or her turn, they turned into the paratroop door and exited the aircraft.

When it was my turn, I stood up, moved to bisect myself in the paratroop door facing the forward portion of the C-130's fuselage. I was the Assistant Jumpmaster and another student beside me was the Primary. We both gave our commands and corresponding hand and arm signals. "Ten minutes!" "Get ready!" "Outboard personnel stand up!" "Inboard personnel stand up!" "Hook up!" "Check static lines!" "Check equipment!" "Sound off for equipment check!" I cleared the door of the aircraft as prescribed. It was my first time ever hanging out the side of an aircraft flying at 1200 feet above ground level. It was simply majestic, and I felt awesome.

Seeing the one-minute reference point, I turned to confirm with the Primary Jumpmaster before giving the one-minute hand signal and call. Then the thirty second reference point over Sicily Drop Zone was parallel with the aircraft. After confirming, I gave the hand and arm signal to the other jumpers. Soon the moment had come. The other student and I locked eyes, nodded, and executed the signal for all jumpers to "Stand-by!"

Then, as if by God's will, the plane started shaking violently due to turbulence. I tried my hardest to avoid moving my feet, but I couldn't help it as I lurched side to side holding onto my static line and trying not go out the door. The black hat grading me helped me to my feet and then the green light was on. I executed a fine exit as I leapt out of the aircraft and into the turbulence of the C-130's prop-blast. No twists or tangles occurred as my canopy opened after a short chaotic drop. I slipped into the wind after checking my canopy and before I knew it, I landed on the sandy terrain of Fort Bragg's largest drop zone, Sicily. PWAC was over for me as I moved to the rendezvous point to turn in my parachute.

After we shook out our parachutes, we returned to the Advanced Airborne School to receive our scores. Lining up, everyone seemed to have passed the exam without difficulty. When my turn came, the Blackhat who graded me noted I performed the exam up to their expectations but that I had committed the fatal flaw of moving my feet. The Blackhat, with another one of his buddies told me that regardless of the turbulence, I should not have moved my feet, even to keep myself from falling.

This sounded completely ridiculous to me as even a two-legged Jumpmaster would of had trouble not swaying in that aircraft. After explaining, the instructor asked if I had any questions. I did not. I turned to walk away angry when he said, "I'm just messing with you, Jumpmaster. You passed!" Of all the lousy pranks. And this was the second one I had fallen for. Breathing a sigh of relief, I walked away a little taller that day. I had become a Jumpmaster.

Graduation occurred with little fanfare. In the classroom, we received our certificates and were reminded that we had to perform the duties of the Jumpmaster within 90 days, or we would need to attend Jumpmaster Refresher. After we were released, I began my two weeks of leave. By now, most of Task Force 2-Fury who had re-deployed were returning from their own leave window. When I returned from leave, I would officially begin my duties as CPT Blackburn's Executive Officer in Brutal Nation. Little did I know this would be the most demanding and rewarding position that I had ever been in.

CHAPTER 18
DRACONIAN MENTORSHIP
NOVEMBER 2014 TO APRIL 2015
FORT BRAGG, NC

I spent my first weeks off from the Army in over a year between Kentucky and North Carolina. Halloween and watching the University of Kentucky vs the Tennessee Vols at Kroger Stadium were the highlights. Most importantly though, Michelle was in full baby mode, planning with her mother Vicki on outfits and decorating the baby's room. She expected a daughter, and I knew she would stop at nothing at getting her way. However, it was still too early in her term to verify the gender. Mom, dad, her father Clay, and Vicki were incredibly supportive in welcoming the next generation into the world.

Back home in North Carolina, we renovated the entire kitchen in our home. Clay, a manager at a cabinet plant in Mount Sterling, helped us redesign the kitchen, choose the right cabinets and features, and drove a U-Haul over eight hours himself for the installation. For three days, I gutted the current kitchen and we prepped and installed our new cabinets. The granite utilized for the counter tops was the real highlight that brought everything together. I felt great getting into home improvement projects and it has become a lifelong hobby. The skills Clay taught me over the years as a master carpenter have become indispensable to me and allowed me to upgrade and modify our homes as Michelle specified. Slowly but surely, my wife was winning the battle in my path towards domestication.

As I was on leave, the battalion continued to roll along. Meetings, synchs, and touch points occurred per the battle rhythm throughout the week. What I failed to grasp while I was gone was that nothing stopped in someone's absence. The Army is designed so that anyone could depart at any time without a major disruption to operations. However, by missing these events, I was falling further and further behind in an officer's most important resource: time.

1LT Ausenbaugh was now the Battalion S4, managing the organization's logistics. CPT Blackburn and our new 1SG, 1SG Andy Anspach were doing their job of leading the unit and my duties in running it. Being on leave while the unit returned to steady state operations put me further behind in achieving situational understanding of Brutal Nation's maintenance, property management, and training resource planning.

Despite this predicament. I was told to not report back in until my leave expired. Since we were still awaiting the return of our overseas equipment, the battalion was just getting back on its feet.

When I did sign back in from leave, my old platoon was now being led by 2LT Adam Ward with a newly promoted SFC Lawson. Adam and I had been high school buddies and we remained in touch throughout the years. Strong as hell and a genuinely good dude, I was more than happy to see that First Platoon was in good hands. 1SG Anspach was a completely new personality than 1SG Goodart, who had recently been reassigned as the Headquarters and Headquarters Company (HHC) 1SG.

1SG Anspach had been in the 3rd Brigade Combat Team and had experience throughout GWOT, Hurricane Katrina, and the earthquake response in Haiti. He had been through countless Joint Regional Training Center (JRTC) rotations at Fort Polk, Louisiana and knew the ins and outs of Fort Bragg from the best training sites, maintenance centers, and facilities where I could obtain supplies and equipment to run the company. In essence, he was one of the best resources a new Company XO could have.

1LT Kamal Wheeler was reassigned to Charlie Company as their XO so we would remain close peers throughout my tenure. His Platoon was taken over by 2LT James Salerno. 1LT Luke Ziller departed the battalion shortly after his redeployment for the Maneuver Captains Career Course at Fort Benning, GA. His successor was 2LT Andrew Giles.

All the Platoon Leaders were new faces straight from IBOLC and Ranger School. SFC Munn of Third Platoon went to take over the scouts at HHC. SFC Wentworth remained the Second Platoon Sergeant. Besides CPT Blackburn, all the officers had rotated out and some of the senior NCOs switched out too. He was rebuilding his leadership from scratch with these young but incredibly motivated officers. The Squad Leaders however, the true strength of Brutal Nation, remained in place for the most part, forming a solid core that ensured we would remain a tight outfit.

I had some new peers in my realm as the XO. `LT Andrew Vogel was the HHC XO, 1LT Nick Luis was the A CO XO, me in BCO, Kamal in C CO, 1LT Thomas Boehm in D CO, and 1LT Katie Munoz in the Battalion Forward Support Company or FSC. Besides our Company Commanders, we now answered to the Battalion XO, MAJ Jonas Anazagasty. Besides Katie, all the other XOs were infantry officers who had served as Platoon Leaders in the companies. Nonetheless, Katie was one of us and we all had the common purpose of running our respective units under the guidance of MAJ Anazagasty. These five company grade officers were about to work harder than they ever had in their lives.

The Battalion Command Team had its share of turnover too, with LTC Zieseniss and CSM Kelly changing out with their incoming successors, LTC Damon Harris and CSM Mcalister. Prior to the change of command, the Battalion Command Team took the officers on a deep-sea fishing trip off the coast as a unique way to have the redeployed officers meet those that had remained stateside or joined the unit during the deployment.

It was an awesome opportunity to meet my future peers and those whom I would work with on a day-to-day basis. The day prior to the ceremony, the Lieutenants all got together inside a ring of PT belts and had to use any means necessary to remain or knock their peers out of it. Kamal, Luke, and I formed a rugby scrum of sorts and started bulldozing the other officers out before we were finally picked off. Third from last man standing, I only vacated the ring when my prosthetic was forcibly yanked from my torso and was physically dragged out. Oh, the good old days of friendly competition. To this day, I believe there is no other person as powerful as a commander on his way out.

Going into November, reset was the main task needed to be completed by all units who had Paratroopers with the re-deployed task force. Our containers of equipment arrived that fall from the Port of Charleston and were dropped behind the COFs by our FSC. All of the equipment had to be unpacked, unwrapped, and restored back into the supply room or sensitive items vault. However, the equipment that remained behind on the rear detachment now needed to be inventoried by CPT Blackburn, along with the forward equipment. After verifying that all of the forward and rear property was accounted for, the separate property books merged back into one. This is not a quick task.

My first duty was to immediately create a plan for the property layouts. The inventory had to account for every piece of equipment, to include the minor components not usually utilized by the end user. Just as we had done in Afghanistan with the change of command inventories, all of it needed to be looked at. CPT Blackburn advised me on how he wanted it done so it went smoothly. End items by LIN order. Then by serial numbered order. The giant layout of equipment should resemble the property book so that all the boss was doing was highlighting serial numbers and marking off pieces of equipment.

This went on until every piece of equipment from a HMMWV to a #10 socket from a mechanics tool set was marked off. As usual, my Supply SGT, SGT Aljamaine Smith was there to assist and mentor me on how to knock it out. This was also when our new Platoon Leaders would sign for their equipment. If it was done right, the Platoon Leaders owned the property in the unit and the command team were signed for none. When CPT Blackburn was satisfied, the Supply Sergeant and I merged the book with the facilitation of the Brigade Property Book Officer.

Throughout the start of this process, other requirements and duties were also completed. We had equipment reset, training resource planning, and maintenance schedules. I had my own office for the first time in my military career. Across from my desk was a giant white board with a list of tasks that needed to be accomplished. As I accomplished a task, I erased the writing and continued. My hope was to have more gaps in the board. To my detriment, however, 1SG Anspach knew there was never a shortage of anything to be done. Anytime he walked by or visited, he'd look at the board and if there was a gap where a task had been completed, he picked up a dry erase marker and wrote up another task that needed to be completed. Thus, there were never any gaps on my board.

The Company XOs had two meetings with the battalion staff on the battle rhythm. On Mondays, we had an afternoon maintenance synch after spending the morning at the unit motor pool dispatching our vehicles. On Wednesday's we had a training and resource synch in the battalion conference room where we planned out all the logistics required to enable a unit to train. In order from HHC to FSC, each XO briefed MAJ Anazagasty and 1LT Ausenbaugh on the unit's maintenance status.

Spreadsheets were printed beforehand that detailed the unit's equipment that was dead lined, how many days it was non mission capable, if parts were ordered or on hand, and when we could expect it to be running again. The training and resource synch was similar in execution, but the XOs briefed the training that was to be conducted, when and where it was to be executed, and the status on logistics such as land ownership and the classes of supply such as food, fuel, construction items, bullets, medical supplies, major end items, and so forth.

MAJ Anazagasty did not like his time wasted. And he disliked poor planning even more. He was smart. He was strong as hell. And he outranked all of us in the room. If one of us briefed something wrong or failed to plan for a critical aspect, he let us have it. Now, he was never an asshole to us. Far from it. Every time we were chastised, it was a learning lesson for the whole group. We were never rated on our performance evaluations by him, our bosses were our Company Commanders. But to ignore his advice, warnings, or guidance would have resulted in painful admonishment which none of us wanted.

But I'll tell you this, all of the Company XO's, the S4, and him were a group where outsiders were not included. He treated us all the same, regardless of who was better and never let us have it in front of someone who wasn't a Company XO. Thus, Andrew, Nick, Kamal, Thomas, Katie, and I grew really close during our tenure because our own success depended on how well we worked together.

Prior to going into the intensive training cycle before next spring's JRTC rotation, we had to have all our equipment that redeployed inspected and fixed by the division and installation's maintenance personnel. Known as reset, our NVGs, weapons, and communication equipment were all checked out and cleared by technicians. Equipment that was non mission capable was evacuated to the Division's Sustainment Brigade for repairs.

As the XO, I had to track what was to be evacuated, physically have it moved to the sustainment brigade's shops, and receive the paperwork indicating that it was no longer with the company. Our weekly maintenance meetings grew longer as more and more equipment was identified to be fixed. However, because of the efforts of the Paratroopers in facilitating this, 2-Fury would be ready to train that winter once we began preparing for war with equipment and weapons that we were confident would function.

In addition to the property maintenance and accountability, I had to manage training resourcing in the company. This was an unending process that began upon the initiation of a WARNO or OPORD and ended after the training was complete in its entirety. CPT Blackburn, based upon the guidance he received from LTC Harris and the operations team, formulated Brutal Nation's training plan that carried the company from the present to JRTC. Working alongside of 1SG Anspach and myself, we then briefed the platoons on what training was on the horizon.

In every plan were the tasks to subordinate units that CPT Blackburn assigned a particular platoon to plan, resource, and execute a unit training event. For four months, we had marksmanship density training, buddy team live fires, squad patrolling and squad live fires. This enabled us to execute platoon patrolling and a company field training exercise (FTX) prior to the culminating exercise, a platoon live fire exercise that was validated by the Battalion Command Team.

My task was to ensure these training events for Brutal Nation had the logistical coverage needed to make the event happen. The level of detail provided by the PL's planning efforts determined how much analysis I needed to conduct. A well thought out plan that was tailored to the training scenario in accordance with the Commander's Intent meant I could simply request the food, fuel, ordinance, and land from the battalion operations team during the weekly training and resource meeting.

A poorly thought-out plan meant I either A: facilitated the planning efforts with the platoon to ensure he understood that logistics drives operations. B: rejected the plan outright and asked the PL to reanalyze the reality of his plan to what needed to accomplish. And finally, C: do it myself since I already had the calculators available.

I always tried to avoid doing it myself when time allowed so that the PL could learn and thus, be a better planner. Sometimes though, when planning timelines were condensed, I had to adjust the logistics concept myself so that the plan succeeded. I quickly learned that a PL's poor plan reflected on CPT Blackburn and myself. Thus, I became the gatekeeper for plans prior to submission to battalion in that I needed to be sure we didn't look like idiots when a PL requested 9,000 rounds of 5.56 green tip for a zero range (true story).

The first month was tough, but as we began training in the field and running more ranges, the concepts of the operations (CONOPs) that were submitted grew better and better. Eventually, they exceeded all our expectations.

At MAJ Anazagasty's direction, our merry band of XO's began a new battle rhythm event. Once a week, we conducted PT as a group. The catch was, PT didn't have to be on base, in a uniform, and could be as creative as necessary. XO PT, as it was officially known, was one of the highlights of my time as Brutal-5. The first session was a running path along the Cape Fear River. On another session, we hit up a Gold's Gym and did spin training on the cycles. Once we even went indoor rock climbing in Fayetteville. But the best session, hands down, was in Southern Pines.

At a small studio called Hot Asana Yoga, we executed PT at 0600 hours and it was something entirely different for me. I had never done yoga prior and with the temperature at 100 degrees, I was completely out of my comfort zone. Nonetheless, the instructor was fantastic and helped me along as I struggled in some positions where having a left leg is beneficial.

After an hour of sweating, we showered and changed for the most important aspect of the offsite PT, breakfast. Our venue of choice was Betsy's Crapes, a short walk away. I am a breakfast man by nature and the selection of meals at Betsy's was singularly outstanding. As we sat around the table, we discussed career paths and opportunities in a low threat environment with MAJ Anazagasty. There was always some type of mentorship going on, even when we least expected it.

During this era, on one cool fall morning, I was unexpectedly invited by the Division Commander, MG Clarke, to attend a breakfast with the Fayetteville Mayor and other prominent locals. Attending with us was another wounded Paratrooper and the Division CSM, CSM Knowles. Our table also had none other than the Commander of the United States Army Special Operations Command, LTG Cleveland. While dining, the other junior wounded Paratrooper and I listened to the conversations held. I did not do much speaking as, due to my rank of 1LT, I was stratospherically outranked, and the discussions held were more political and strategically focused than what I had been educated or involved in.

LTG Cleveland, after hearing a little more about the guests at the table, asked me about my interests in special operations. As if cued, CSM Knowles asked me, "Didn't you submit a packet for SFAS?" To this day, I still have no idea how he knew. Intrigued, LTG Cleveland asked before I could reply when I had a date for the course. I politely informed the LTG that I did not, I was told that the course couldn't accommodate an amputee. "Well, I can fix that," LTG Cleveland informed me, matter-of-factly. "We can get you in the next selection date without issue. What do you say?"

It sounded like a dream come true. I could get a fair shot at becoming a Green Beret, my lifelong aspiration. I had a champion who had my back. However, something in my heart was holding me back from committing. The Commanding General and Command Sergeants Major of the 82nd Airborne Division had been instrumental with my chain of command in getting me back on active duty. To jump ship after everything they had done to lobby for my return seemed like blatant desertion. I also considered for those brief seconds of reflection if I would truly be given a fair shot in the Special Forces Qualification Course if I needed the Commanding General himself to sponsor me. This would not have been a secret and all of the cadre, and my peers would know. My pride was immensely shaken when my packet was declined.

I looked LTG Cleveland in the eyes and could feel the eyes of the

Division Command Team on me. I politely declined his offer, citing that I felt I would be of better use to the Army commanding a company of Soldiers. And just like that, I ended my final chance at ever becoming a Green Beret.

Later on, during the return trip to the Division Headquarters, CSM Knowles told me that I had made the right decision in front of his boss and that my loyalty wasn't unnoticed. I received both his coin, and that of the Commanding General. To this day, I reflect on that morning because a single decision altered my military career. And most importantly, I made it on my terms.

The day after Halloween is an important date in the Pitcher household, according to my wife Michelle. On November 1st, I am tasked with dragging down boxes from a storage room or attic and their contents laid out for inspection. Throughout the day, a half dozen artificial firs and spruces are assembled of various heights and strategically yet tastefully emplaced throughout our home. A large tree in the sitting room, a smaller tree in the kitchen, a medium tree in the dining room, and even a small tree in my office. The tallest tree was reserved for the living room. Each Christmas tree had its own unique theme. A colorful tree of every pastel design, a red and green tree, and an assortment of other patterns. She collected the ribbon and specialty ornaments for each specific tree going back all the way from when we first started dating in 2008.

Her mother's Christmas trees resemble something Martha Stewart would envy, and Michelle competed every year to outdo Vicki. The result was nothing short of spectacular. The flourishes, the mantle pieces, and the display in the foyer made the interior of our home something special. Even the presents received the royal treatment. If you think you could sneak a present under her tree you are in for a rude awakening. Even relocating or removing an ornament would incur her wrath. She wrapped every single gift with paper, ribbon, and bows that complemented the tree they were under. Each set of presents echoed the theme of the tree that overshadowed them.

My job was much easier. I had to give the trees compliments every time I passed by them. And my main responsibility was to put up the Christmas lights on our home. Now, compared to her efforts, my light display was mediocre at best. But compared to the neighbors, it was still one of the brightest homes in the neighborhood. Thus, we initiated the holidays in 2014 with a grand display. This was to be our first Christmas together in our home, so we went all out.

Ironically though, Christmas Eve was spent at my parent's home and Christmas Day was spent at hers back in Kentucky. After all the work, we didn't even get to see our home over Christmas. But we still had two months prior to Jesus's birthday to enjoy it, so I guess that went for something.

Prior to the holiday exodus, we were immersed in getting ready for collective training. All the companies conducted marksmanship density throughout December before the weather turned for the worse. The XOs ensured the platoons running the ranges had everything they needed and, once the training initiated, returned to the office. For me, there was still a lot of issues with the property accountability piece. Almost all the shortage annexes we had on file were outdated or inaccurate. CPT Blackburn asked me to input a system to fix our property accountability for the next change of command.

Much of my free time now was spent combing through technical manuals of the components of the end items and the basic issued items of every end piece of equipment. I then created or updated the corresponding shortage annex, listing out the pieces of equipment by type and quantity. The result was a more accurate way for the officers and NCOs to execute inventories during monthly and cyclic inventories.

This process took months due to the sheer number of items on a property book in a rifle company. HHC and FSC had even more, so I considered myself lucky. Some items had only one page, such as a HM-

MWV. Others, such as the mechanics tool sets, became the bane of my existence with dozens of pages due to having to account for every single wrench, socket, drill bit, etc. While I was not doing the "hooah" stuff that people typically think about when visualizing Army training, I was building combat power through improving the organizational systems in place that resulted in CPT Blackburn being able to focus on leading Brutal Nation and not on the tedious aspects of the unit.

It was in November when I received news that completely changed my career path. My nomination packet for the United States Marine Corps Expeditionary Warfare School (EWS) at Quantico, Virginia had been selected. It probably helped I had letters of endorsement from my Company Commander, Battalion Commander, the United States Africa Command Commanding General, General David Rodriguez, and the Senate Majority Leader, Mitch McConnell. With this news, I was no longer going to permanently move from Fort Bragg, North Carolina to Fort Benning the following summer. We planned to keep our home at Bragg and rent a home in the vicinity of the Marine Corps Base only four and a half hours away so that we could continue seeing our friends in Fayetteville on days or weekends off.

It was no surprise that 1LT Doug Ausenbaugh, the Battalion S4 who worked directly for MAJ Anazagasty and facilitated the Company XO gang's efforts, was also accepted into the course. Of the hundreds of packets, only eight were selected and two came from 2-Fury. I still had ten months prior to the commencement of the course next August so there was little need to be worked up too much about it. I was too busy facilitating CPT Blackburn and 1SG Anspach's drive towards getting Brutal Nation ready for JRTC next spring, so EWS just became a pin on the calendar for the time being.

I began executing my first Jumpmaster duties that winter. One of the best aspects of being a Jumpmaster of any rating is that you rig up in the harness after everyone else has been inspected. For some of the more seasoned Jumpmasters, rigging up in the harness occurred only moments before station time and loading the aircraft. Just as in the course, I JMPI'd Paratroopers in my chalk. Assigned the duties of the safety for my first duty, I facilitated the chalks Primary and Assistant Jumpmasters' duties in preparing the Paratroopers to board the aircraft, emplace them in their seats, and help the Jumpmaster team prepare the aircraft for the airborne operation. It was a lot of work but luckily, the duties were a carbon copy of the schoolhouse, so I felt fully prepared.

I felt a matter of pride for every Paratrooper I inspected. When station time came, the chalks began rising from the special benches constructed at Green Ramp that were modified to allow Paratroopers to sit while in the parachute harness. Primary Jumpmasters shouted one command for their chalk to signal the movement to the aircraft. "Chalk 101 on your feet!" "Chalk 102 on your feet." The chalks of roughly 100 Paratroopers picked up to move.

We were doing a night airborne operation over Sicily Drop Zone by a C-17. 100 heavily laden Paratroopers rose to their feet and executed a left or right turn to face the giant garage door facing the apron where the aircraft parked.

I pulled a collapsable storage container full of supplies necessary for the operation. Moving up the ramp into the cavernous maw of the C-17, I parked my container and began helping the jumpers get into their seat on the starboard side of the fuselage and buckle in. I passed water bottles around to the jumpers to remain hydrated. Once everyone was loaded and seated, I took my place next to the Primary Jumpmaster and hung out as the tail ramp closed off our access to the outside world apart from a small peep hole on the paratroop door.

Soon we were aloft over the Cumberland Country area. It was a smooth flight and time elapsed quickly. Suddenly, the Air Force Loadmaster informed the Primary and Alternate Jumpmaster that we were within ten minutes of the drop zone. I put on my parachute, different for safeties than jumpers because it wasn't hooked up to an anchor line cable. In the event I needed to exit the aircraft for an emergency or fell out, it was up to me to activate my own canopy. It was much lighter than a T-11 and it did not have a reserve parachute hindering my movement.

"Ten Minutes!" shouted the Jumpmaster team in unison. I moved to the bow of the aircraft along with my fellow safety, preparing myself to inspect static lines back to the paratroop door. "Get Ready!" The jumpers prepared for the immediate "Outboard personnel stand up!" 50 Paratroopers rose to their feet nearest to the skin of the C-17 Globemaster. "Inboard personnel stand up!" The remaining 50 Paratroopers facing the outboard portion now stood up and interlocked. "Hook up!" The jingling of static line snap hooks connecting to the anchor line cable briefly took over the roar of the engines. "Check static lines!" The jumper checked to ensure their static line snap hook was properly fastened in the right direction and that their static line was properly routed to ensure it didn't choke oneself after departing from the aircraft. "Check Equipment!" All Paratroopers conducted an inspection of themselves to ensure they were ready to go, all points of connection being properly fastened. The jumper immediately in front was also inspected from behind by their fellow jumper, a level of trust needed by all in their fellow Paratrooper's training.

At "Sound off for equipment check!" the Paratroopers began slapping rumps and shouting "Okay!" This was my cue. I began moving to the first jumper inspecting all the static lines and ensuring hands grabbed the static line in a way to create a bite for me to grab. I checked the snap hook, hand position, and traced the yellow cord back to the pack tray. I did this quickly, covering fifty jumpers as fast as I could. I heard "All okay, Jumpmaster!" This signaled that the entire echo had reached the number one jumper who slapped the hand of the Primary or Alternate Jumpmaster. Soon, I reached my counterpart, and we were ready to rock and roll.

I watched the Primary Jumpmaster hookup and immediately inspected his static line and anchor line cable. The Loadmasters opened the doors to the aircraft, cold air from the pitch darkness now filling the interior. The sound was deafening. "Army, your door!" The Loadmasters handed off the paratroop door to the Jumpmasters who inspected the doorway, the door itself, the wind deflector, and the jump platform. The entire time I held the static line and assisted in the inspection by ensuring he was free of any obstructions as he stepped out onto the platform. Once he was satisfied, he began scanning the horizon for the one minute and 30 second reference points. "One minute!" I was tensing up. I was about to rake fifty static lines as jumpers exited. "Thirty seconds!" I gave the Primary a bit of room as he turned back into the aircraft and eyed the Assistant Jumpmaster. With a synchronized motion, they roared "Standby!"

The number one jumper stepped up and was moved into the paratroop door. With one swipe of my arm, I secured his static line and pushed it aside. Time stood still. My heartbeat pounded as I watched the jumper, the Jumpmaster, and the light. The jump lights, currently amber, suddenly turned green. "Green light, go!" The number one jumper disappeared into the void. Then the second. Then the third. Their pace picked up as they exited the aircraft. I swiped static lines from each jumper and pushed them behind the Primary. I watched the light out of my peripheral, prepared to halt the jumpers in event the red light came on. Thankfully though, it was over quickly. Once the last jumper exited, the Primary Jumpmaster watched as his Assistant exited. Once he was gone, my counterpart departed for his jump, his duties in the aircraft now complete.

Once all jumpers had departed, we began pulling the static lines back into the aircraft. I never realized just how much force was pulling on those static lines as I needed every ounce of my strength to pull them in. Typically, the Loadmaster can just reel them in with a pneumatic winch, but we did not have this option. I was sweating profusely as the last of the static lines were pulled in. Each Loadmaster was assisting us in the process. I would have felt bad, but the other safety was having similar issues as well. 50 static lines and pack trays aren't exactly light.

Once the final static lines were pulled in, the paratroop doors were shut. The worst part of being a safety happens at this point. The static lines, swaying in the chaos of the aircraft's slipstream were now a giant tangled mass of yellow spaghetti. For the remainder of the flight, I snaked individual static lines out and bundled them into aviator's kit bags that we had brought aboard just for that purpose. As the aircraft came in for its final approach at Pope Army Airfield, I was still separating the mass. After pausing to land, I completed the remainder as we taxied to the apron in front of Green Ramp. The other safety and I packed up our gear and headed to the Departure Airfield Control Officer (DACO) to be released. Prior to stepping off the aircraft, I swapped patches with the loadie, my first of many to be collected.

Nobody had been hurt during the operation. All Paratroopers had exited over Sicily Drop Zone. We made station time and had not lost anything. All in all, it was a textbook airborne operation from our perspective. There were many more ships in flight dropping jumpers, so we didn't know how the other chalks fared. But to me, I was proud. I was a Jumpmaster in the 82nd Airborne Division and representing 2-Fury and Brutal Nation. This one event sparked my lifelong love of Jumpmaster duties and I volunteered for as many as possible for the remainder of my time on jump status.

Just prior to the long-awaited Christmas exodus, I learned one more lesson from MAJ Anazagasty. One of the issues identified after 2-Fury relocated to its new digs from the old 4th Brigade Combat Team footprint was that all of the arm's rooms in the companies needed certification to store ammunition. The structure of the vaults themselves was sound, but a checklist of nuances needed to be completed prior to the certification to store Class V ordinance by the proper entity. The checklists for every unit were more or less the same and the task for the certification of the arm's rooms generally fell to the XO team.

As I scanned the lengthy lists of requirements, I could tell we had almost none of the documents and signage that amounted for most of the tasks. About a dozen memorandums were needed. Chains needed to be on the cages. Fire suppression systems needed to be operational, signage for hazardous materials needed to be installed inside and outside the vault, inside the COF, and outside the COF. Vault personnel needed to be vetted, listed on access rosters, and a handful of technicians and employees on Fort Bragg needed to inspect the premises altogether. The XO team quickly put our heads together. Who had memos we could copy? What chains were needed? What signage was authorized. We identified how we could tackle this. 1SG Anspach was essential to this endeavor. He took me to the base supply store which had the signs I needed and showed me exactly how the vault needed to be configured to get the Fire Marshall's blessing.

For almost a week, 1SG and I worked to check off the checklist. Time was of the essence to get the vault certified to store ammunition as we began range density operations soon. Failure meant the company had to store ammunition at field ammunition supply points. This entailed an armed guard 24 hours a day, seven days a week until the ammunition dunnage was turned back in. With our vault certified, the ammunition could be stored inside and secured. No guard needed and nobody was standing outside in the fast-approaching winter season. In essence, getting the vault certified for ammunition storage was taking care of Paratroopers.

Once the memorandums were signed by CPT Blackburn, they were posted where policy and regulation demanded. The lockers in the vault were secured to the wall with galvanized steel chains. Signs indicating that the building stored Class V were mounted. Fire inspections, facility inspections, and operations security and force protection inspections were carried out. With only days prior to the unit's draw of ammunition and ordinance for marksmanship density week, we received the final form stating that we were certified. It was a big win for Brutal Nation.

However, during the next Training Resource Meeting, the issue of the vaults being able to secure ordinance came up and with it, the status of our arm's rooms. As MAJ Anazagasty inquired, I triumphantly announced that we were good to go. Instead of praise, however, I received a question. "What did you do to facilitate the other XO's in their efforts?" I was silent. He was absolutely right. I had focused so much on my own unit I failed to see the bigger picture. Brutal Company was but one of three rifle companies and one of six for the whole battalion. The lesson was simple for the BN XO without him needing to say it. Either we succeed as a collective whole or fail together.

Luckily enough, my peers had cracked the code too. One by one the remainder of the arm's rooms across the battalion were certified for training. We had not realized it yet, but our behind-the-scenes efforts were generating significant combat power for 2-Fury that would pay dividends the upcoming year.

It was during this time that Michelle was finally growing with our child. Just prior to leaving for Kentucky for two weeks of holiday leave, we went to an ultrasound clinic to determine the gender. In her mind, she was having a daughter. I'm pretty sure she had outfits, furniture, and toys already picked out, saved in an online shopping cart, and awaiting the final go ahead to push pay.

At the clinic, her world was turned upside down when the technician showed the blurry image of our baby. With two clicks on a spot on the screen, she announced that we were having a boy. I could have laughed. After all, my sibling was a boy, my cousins were boys, and my grandfather's children were all sons. Riding in the car to dinner, Michelle wasted no time calling her mom, Vicki to give her the news. "Agh! It's a boy!" was how the conversation started. Immediately, she was given the scolding of her life by my mother-in-law. I wasn't privy to the conversation but from what I gather, she imparted some hard wisdom on her daughter. At the end, Michelle apologized and cut off.

She apologized to me and said she should have been thankful that we had a healthy baby boy on the way. I felt bad though. I know in her soul she wanted her own daughter and that a boy was not in her plan.

Despite the truth, she switched gears immediately towards preparing for a son. It took no time at all to decide on a name. We debated between Hunter, Corbin, and Connor but decided on the latter. And so it was, our first child, our son, would be known as Connor Scott Pitcher, with an expected delivery date of late April of 2015.

December at Fort Bragg ushered in the much-anticipated airborne operation put on by the U.S. Army Civil Affairs. Known as "Toy Drop," Paratroopers all over Bragg participated in a lottery to earn a coveted set of foreign jump wings from whatever nations participated during that year. They "bought" their lottery ticket by donating a toy to the local community at the entrance to the pax shed at Green Ramp. Any new toy, regardless of cost or size, ensured a ticket. Only one ticket could be issued per jumper but there was no limit to the number of toys one could donate. If I recall correctly, units at Fort Bragg were allotted a number of hard slots or "chutes" per battalion where Paratroopers could jump for foreign wings as a retention incentive, being an outstanding Paratrooper, or in some cases, being a pay loss for not having jumped within 90 days.

On a cold Thursday morning, I showed up to Green Ramp with a small girl's bicycle to donate. The line stretched around the PAX Shed and well into the parking lot, thousands of Paratroopers waiting for just the few hundred slots available on a manifest. I bought the bike from the local Toys-R-Us in Fayetteville and decided to give it the old college try. The NCOs in 2-Fury had told me the rumor was that a gaming system like an Xbox or PlayStation or a bicycle was a sure-fire way to get one's number called on the lottery. I decided to test that theory.

After obtaining my ticket, I waited in the PAX shed. I already had Chilean foreign wings, so I wasn't too concerned about my ticket number being called. After four hours of hearing numbers called, I donated my ticket to one of my guys from the Brutal Company Headquarters and left. I had a lot to do at the office and ranges were still being conducted that day. Later on, I found out that my donated ticket's number had been called and he was placed in a chalk.

That afternoon, somebody told me to try and strap hang on Saturday morning. I thought that would be insane as everyone wanted to be on one of those drops. However, wisdom was parted on me by the NCO who told me that the Final Manifest Call (FMC) at Sicily Drop Zone was at 0300 hours on Saturday morning. Not only was it a day off but it was well before the start of a typical duty day. He explained to me that the vast majority of jumpers are junior enlisted Paratroopers who will probably be out drinking and hanging with friends the evening prior. Odds are that some would not wake up in time for FMC. I decided to give it a shot.

I woke up at 0130 hours on Saturday, careful to not disturb my pregnant wife. I drove the 45-minute drive from my house to Sicily Drop Zone, the streets along Longstreet and Manchester being almost deserted. I parked in the lot where everyone else had and wandered over to what looked like a command-and-control tent. I introduced myself to someone from Civil Affairs with a clip board and asked if there were any openings. He said something along the lines of whether or not I donated a toy and I concluded that I had indeed brought a bike to the lottery.

What sealed the deal, however, was when I mentioned I was a Jumpmaster. All of the jumpers needed JMPI'd so they needed as many Jumpmasters as they could get. I was placed on a chalk for a pair of Royal Netherland's Parachutist Wings as the number one jumper. It didn't get much sweeter than that.

That morning, I JMPI'd about twelve jumpers and rigged up for the jump just prior to station time. It was a C-130 jump, but the ceiling or cloud cover was too low for most operations. However, this aircraft was equipped with the modern Adverse Weather Aerial Delivery System (AWADS) so this jump was happening. After a short flight, I was standing in the door facing only a white cloud. When the Dutch Jumpmaster completed his duties in the aircraft, I waited for the green light. Once out the door, I fell through the clouds, unsure of what was going on. Although my canopy opened without incident, I could not compare my rate of decent to any of my fellow jumpers. After what seemed like an eternity floating down through the ceiling, the ground appeared, along with the other parachutes floating down to earth. Never before had I had such a soft landing.

When I turned in my chute, I saw my old friend 2LT Adam Ward. He also had a good jump and was also enjoying his day off by earning a sweet new set of jump wings. The caretaker of my old platoon, many of my old boys were also jumping so it was a short reunion of sorts. I didn't stick around too long though. They were now his team and I had to remember to respect that.

2015

My leave closed out 2014 and ushered in 2015 with minimal fanfare. Apart from expecting a spring baby, Michelle and I made the rounds seeing families and friends prior to returning to our home in northern Fayetteville in early January. The schedule for the next four months was hectic. Brutal Nation was essentially in the field with weekends off for the rest of winter and well into spring. We had squad patrolling for a week. Then squad live fires for another week at West McKeithan's Pond Training Area. Then the battalion was holding platoon patrolling prior to a platoon live fire exercise at OP-13. The unit would then transition to outload and deploying to Fort Polk for a rotation at the Joint Readiness Training Center (JRTC) in April.

The training areas north of the Fort Bragg cantonment area were our area of operations for the first week of squad patrolling or STX lanes. The Company CP was set up by me, our supply team, and the Paratroopers working in the company operations section that assisted 1SG Anspach. We had a HMMWV with a small trailer to carry us out to a remote training area. In the cold and overcast conditions, a small general purpose or GP Tent was put up along with a COM-201 antenna high in the trees. The platoon's occupied bivouac sites near us and their squads would patrol on lanes developed by CPT Blackburn and his planning OIC.

My job was straight forward. Provide logistics to sustain operations and, in the absence of the Commander, direct the training for the squads. Luckily for me, CPT Blackburn was in his zone when out in the field and relished every opportunity to train alongside the Paratroopers. Our supply team led by SGT Aljamaine Smith ensured the platoon's had what they needed and coordinated with the Platoon Sergeants for hot chow pickup from the DFAC in the mornings and evenings. In the winter, hot food makes the difference for the team. And whenever we could obtain it, we strove to do so.

Once steady state was achieved and the systems such as chow resupply and medical evacuation were in place, my place of duty then reverted back to the office. I was still working the property book shortage annexes while the unit was in the field. I became obsessed with getting every piece of equipment documented correctly and preparing for JRTC. While CPT Blackburn advised me to go home in the evenings to my pregnant wife, I refused due to the fact the Paratroopers could not do so likewise. He was taking care of me by continually reminding me that my wife needed me at her side. However, I saw my place either in the field or in the office during the duty week. I confronted myself with knowing I would have the entirety of this fall and next spring with my wife while learning the operational art and design of the Marines.

On the last day of the SQD STX lanes, I bought around fifty pizzas from Little Caeser's. It had been a miserable cold week for those boys, and I wanted to do something for them when they returned. While a return from a field problem meant a brief respite for the Paratroopers; for my team, it meant ensuring all equipment, weapons, and vehicles were accounted for, turned in, and anything broken documented and prepared for evacuation to the Brigade Support Battalion.

Running estimates had to be updated to show the Commander and

1SG that we still had enough combat power for the unit's culminating training evolution at company echelon, the squad live fires at West McKeithan's Pond. Battalion's train and validate platoons. Companies do the same with squads. In order for us to be ready for the battalion's validation of our platoons, it was imperative that we ensure our own squads were validated in a live fire exercise.

As I devoted myself to my career, Michelle dedicated herself to what she does best, making a house into a home. And in this case, preparing 7630 Trappers Road for the arrival of baby Connor. I remember walking up the stairs to see that she had taken the empty bedroom that we had always reserved and completely transformed it. An entire wall was painted, with a navy-blue stripe running across with huge wooden white letters spelling out CONNOR. Below his name was a white crib and around the room were all the other fixtures such as a diaper changing table above a matching dresser and a comfortable rocker for mother and son. I adored how Michelle prepared our home for Connor. She did so with the same amount of love, devotion, and attention to detail as she had when setting up for Christmas.

During this time, I sold my Harley Davidson. Not because I needed the money but as a precaution. I had laid it down on the way to Jumpmaster School one morning and endured only scratches and a bruised ego. I was lucky but was also foolish enough to inform my wife of my mishap. Her concern and anxiety every time I rode to work only worsened. I put a lot of work into that bike, with a new black and brass springer front end and new ape hangers. It was a beauty and I loved it. When someone offered me a deal that I couldn't refuse for it, I sold her. I may now be without a bike but at least Michelle rested knowing I wouldn't be another statistic on the road.

Life as a Paratrooper in the 82nd Airborne Division is busy and fast paced. It is hard on the families and body. But one thing it makes you realize is that the time that you do have with family is precious. My few days off between field problems were spent with Michelle and my Boykin Spaniel that winter healing the wounds caused by the previous year's distance. But more importantly, it got us in the right place to bring our son into this world. Give him assurance that his parents, his mom, and dad, would always love him and be there for him.

The Company executed its SQD LFX at WMP without incident. Just like the Squad STX, my job was to get the training area set up, the command post established, and to ensure the logistics were taken care of. Once everything on the checklist was crossed off and the squads were ready to go, my initial work was done. I could do more for the team at the office knocking out projects and making speedball runs of energy drinks to the team.

CPT Blackburn and 1SG Anspach were getting their miles in this week. Nine rifle squads needed to be validated. Each squad had to execute a day dry, a day blank, and a day live iteration prior to conducting a night blank and a night live run. Thus, each squad executed the training lane five times totaling 15 for the platoon leadership and 45 total runs for CPT Blackburn and 1SG Anspach. Just like SQD STX, pizza was awaiting the team once they returned to the COF on that Friday.

After Brutal Nation had validated their squads at echelon, it was the 2-Fury Command Team's turn to validate the platoons. The operations team put together a plan for all nine rifle platoons and the heavy weapon's sections to patrol in STX lanes. Concurrent to the STX lanes was a platoon live fire exercise at OP-13. It was during this time in late February and into March that we seriously began planning our outload for JRTC.

Prior to the training, the Battalion Command Team and staff had already spent a week at Fort Polk for the Leader Training Program on executing the Military Decision-Making Process. From the rumors I heard, it was a giant party for them, but I couldn't confirm just what happened in the off-duty hours. Within a month though, it would be my turn to head to Louisiana. The XO's and Supply Sergeants from the unit's would be first to go to JRTC to set conditions as the White Cell and Torch Party would receive the unit and conduct reception, staging, and onward integration (RSOI). MAJ Anazagasty was leading the White Cell and rest assured, our time would be spent wisely and efficiently.

I was going to have an interesting experience at JRTC. I had never been to one previously, but from what I gathered from all who had witnessed a rotation, it was universally hated. The battalion would link up at an Intermediate Staging Base (ISB) at Alexandria Airport near Fort Polk and execute a Joint Force Entry (JFE) into the "box" or battle space by airborne assault. Michelle was due to give birth during the rotation so I would only be going to Louisiana to set conditions for the arrival of Brutal Nation as a member of the White Cell. The other Company XOs and I, MAJ Anazagasty, the Supply NCOs, and 1LT Ausenbaugh would fly in early to Fort Polk and get things squared away. A couple of days prior to the battalion's airborne assault, I would fly commercially back to Fayetteville and be ready to receive my son into this world.

That winter, my Supply NCO, SGT Aljamaine Smith was reassigned to another unit. There was no backfill, much to my frustration. 92Ys don't exactly drop off trees like crab apples, so obtaining another Supply NCO was not in the cards. His assistant was SPC Jean, a good Paratrooper of Haitian decent who was woefully unprepared to outload the unit for JRTC. I had to mentor him the best I could. English was not his first or second language, speaking Haitian and French first.

Jean worked hard but the communication barriers were huge. In one episode, the battalion was scrambling to find knuckles to bind shipping containers prior to their movement to Fort Polk. I asked Jean if we had any of these giant egg-shaped iron pieces to which I was surprised to learn that we in fact had over 80 sets. I immediately informed 1LT Ausenbaugh and the other XOs of this information. Upon inspection later on however, I discovered the "knuckles" were little bags of nuts and bolts. So much for the windfall.

One time, SPC Jean ordered the entire company brand new Oakley

M Frame eye protection, but didn't tell a soul about it. Although the boys would love that, the fiscal management process headed by MAJ Anazagasty was not about to have it. I remember being called into the Battalion Commander's office with CPT Blackburn and having to explain it. CPT Blackburn graciously explained how SPC Jean's purchase order was made without approval and that this helped justify why we need a seasoned Supply Sergeant as soon as possible.

Ironically enough, Jean would be commended by the Battalion Command Team for looking after the welfare and safety of the Paratrooper's eyes. Luckily, enough were ordered so some of the staff and command team could benefit too and the issue was dropped.

The final humorous tale of SPC Jean was when we were loading out the company for JRTC. The Quadcons were behind the COF, and we had a limited amount of space for all of the weapons, equipment, and supplies needed for the company to fight and sustain itself throughout the rotation. I made a packing list and delegated the load out to Jean. Coming back later to supervise, I witnessed that half of the small container was packed with boxes of government issued toilet paper. The kind that is so thin that it is almost translucent, but as rough as sandpaper. I couldn't believe it. Jean truly packed enough toilet paper for the entire company for a year and believed this was the priority.

We unpacked it together and repacked it in accordance with the prescribed list. SPC Jean, one of the politest Paratroopers I have ever met, truly cared about the boys in his heart and his stories of voodoo practices still haunt me to this day. Apparently to keep demons away, Jean had informed me that I should open a can of coke and put a banana in a tree. I still have not tried it.

Michelle's doctors' appointments showed that Connor was healthy and developing on schedule. Her co-workers at the school she worked for as a teacher threw her a massive baby shower and Connor's room began to overflow with gifts from family, friends, and neighbors. I was thankful that I was not deployed for this momentous occasion for our growing family. I was even more thankful that my leadership in 2-Fury cared enough to ensure I didn't miss my son's birth.

I would rotate out with another 1LT who would be taking over as the official XO of Brutal Nation, 1LT Chris Yackley. 1SG Anspach was also switching out before the rotation, being reassigned to Louisiana State University Army ROTC. SFC Willie Wentworth III, the Second Platoon Sergeant was taking over for 1SG Anspach. CPT Blackburn and SFC Wentworth worked well together so I knew the team would carry on just as successful as before.

About ten days prior to the JFE, the White Cell party flew to Alexandria Airport and drove to Fort Polk. We would spend the first week of our stay at Tiger Land in north Fort Polk getting the sustainment accounts squared away and receiving containers, vehicles, and follow-on personnel. Our duty days were long and there was no shortage of things to do. XO PT now occurred every day with long runs or time in the gym.

Perhaps my favorite memory though, is going to grab bags of boiled crawfish, potatoes, shrimp, corn, and sausage from the food truck outside of the PX to chow on as we worked. I had lived at Fort Polk from 1999-2000 so much had changed, but the general layout remained unchanged enough that I could help orient the others.

After the vehicles arrived by train to the post railyard, the drivers who had flown in on the ADVON party took custody of the vehicles and moved them to be fitted with the equipment needed in the box. Afterwards began the long convoy to Alexandria. Luckily for me though, I rode in the GSA van instead of a tactical vehicle. Unluckily for me though, I was now staying at the ISB until I flew home.

If there was a place that can compare to a gulag, it would be the ISB. For starters, back in 2015, there was no barracks or tents. Only pavilions for the units to sleep under. The XO team signed for the entire battalion's requirement for cots and spent a good portion of a day setting them up under the pavilions. Thankfully, the weather was warm and sunny for the most part. The food was abysmal, worse than anything I ate in my career. And you were surrounded by barb wired fences. In essence, it was a miserable place and I was glad the boys were only staying here briefly.

A couple of days prior to the jump, the main bodies began to arrive at the ISB. I immediately met up with CPT Blackburn and the PLs to give them a layout of the land. They looked just as unhappy as I was with the current digs upon first inspection. Our motor pool was a field separated by a snake infested creek with a bridge spanning the water. Vehicles were parked in line by type and unit. Brutal Nation wasted no time getting them prepared for the training. Tough boxes and supplies were stacked in the LMTVs while camo nets and black out tape went over the HMMWVs and every light. I did my handoff with 1LT Yackley at this point, having him inspect and know where everything was. I did, however, fail to remember to have him sign for an M2 .50 Caliber Heavy Machine Gun Blank Firing Adaptor (BFA) which would later bite me in the ass when the unit was clearing Fort Polk. This oversight ended up costing me $500.00 a month later. Lesson learned, don't have anything signed over to you in the field.

The night prior to my departure, the XOs did a group pic at the Battalion CP. I talked at length with MAJ Anazagasty for career advice and feedback, which he graciously provided. I said my farewells to the team. It had been a great run and without a doubt, being a Rifle Company Executive Officer was one of the hardest and most rewarding positions I had ever been in. CPT Blackburn wished me the best and to bring my son by when I was off paternity leave. Indeed, he left his mark on the organization. The culture he created was singularly exemplary regarding command climate and readiness.

Josh and the XO Team of 2Fury at the Joint Readiness Training Center,
(APR 2015)

CHAPTER 19
THE CON MAN
APRIL 2015 TO JULY 2015
FAYETTEVILLE, N.C.

The next morning I flew home to Fayetteville to be with Michelle. She was the main effort for the rest of my life, and I had to prioritize her and our son in everything I did. The days between my homecoming and her inducement were long. We waited for any sign that Connor was on his way. The clever girl Michelle is, she looked up ways to get her vitals up in order to convince the health care providers that she needed to be induced. After almost nine months, she was done being pregnant and wanted to see Connor.

After running up and down the stair's multiple times, we went to a regularly scheduled appointment to check up on her status. With her blood pressure elevated, the physician made the determination to admit her into Womack Army Community Hospital for inducement due to possible preeclampsia. She had fooled everyone. She even had her hospital bag in the car ready to go, complete with Connor's onesies for the photos she was sure to take.

The time was now. I called CPT Blackburn and SFC Lawson to let them know Connor was on the way. I called my parents. I called everyone. And we waited. For three days. Connor was already displaying a fighting spirit prior to his arrival. Michelle's plan was for Connor to come out on her terms, but that was not the case. In the early morning of April 22, after over 36 hours of labor, Connor Scott Pitcher finally arrived in this world, weighing a healthy 7 pounds 4 ounces. He had my heart immediately and was perfect in every way, save for a little jaundice that quickly cleared up.

Michelle wasted no time in getting him into his newborn onesies for the pictures. She could barely move, but she was obviously very much in control. I could only imagine what the baby would be dressed in had it been a daughter. After a day in the hospital recovering, we were evicted back to our home. Aussie immediately bonded with her new brother and Connor seemed to likewise return the affection. I was only twenty-four years old and was now a daddy. It was a feeling I had never felt, to be someone's complete world whose very existence and livelihood depended on you. I spent most of my last weeks in 2-Fury on leave prior to leaving for Quantico learning the ropes of fatherhood. It wasn't too bad when I reflect on it. Connor slept an astounding 12 hours from 1900 hours to 0700 hours routinely and had no issues eating. He was quiet and only fussed when he needed a change or was hungry.

My parents and Michelle's own came to visit their new grandson. Michelle, still healing, relied on me to do much for Connor, but I was clueless on some of the other aspects. The assistance from our parents and from our neighbors was immensely helpful. I mean, is anybody ready for parenthood? We were so blessed by the shows of affection and love. Dinners, homemade blankets, and gift cards flowed in from well-wishers, and everyone wanted to see baby Connor. I was just thankful that I would be home for much of his first year, locked in at the schoolhouse instead of on a deployment rotation to the Middle East.

On May 1, just after 2-Fury returned from their successful rotation at JRTC, I was promoted near the Falcon Brigade Headquarters by the Battalion Command Team. Now a Captain, after being in the U.S.

Army for four years, I earned just enough more to be able to afford rent in Dumfries, VA, just outside of Quantico. In this method, I could afford my mortgage and still live outside of the Marine Corps Base.

The best aspect though, was knowing that I had spent three of those four years as a living breathing example of just what a human being could accomplish through sheer willpower. I lost my leg on April 15, 2012 to a Taliban IED, just eleven months after commissioning as a 2LT from EKU. Through a rigorous training regimen of occupational and physical therapy and the facilitation through my healthcare providers, family, and friends, I beat the odds and returned back to jump status in the 82nd Airborne Division. I earned my Expert Infantry Badge, deployed for a second tour to Afghanistan, and earned the title of Jumpmaster. Now, perhaps the most important accomplishment of them all, I became a father, when physicians cautioned that this noble calling may not be in the cards due to my injuries.

I must remain humble though, as this was not completely on my own. The leadership in the division and 2-Fury was instrumental in ensuring my success. They gave me every chance and never lowered the standards for me. I have had a chip on my shoulder for three years, yet they treated me like one of their own. My peers, the PLs and XOs I served with, were my support network for the last two years, enduring hardship and success together and forging those strong bonds of friendship. The NCOs and enlisted, my Team Leaders, Squad Leaders, Platoon Sergeants, First Sergeants, and Command Sergeants Major who guided me, mentored me, and yelled at me every step of the way. Without these incredible and selfless people, I would have failed.

But most important of them all, my wife Michelle, who endured more than everyone else. Who was told I was dying after getting the initial phone call? Who stopped her studies for her graduation finals to come to my bedside at Walter Reed? Who has put up with more bullshit than anyone else? Who married me only to see me deploy again to the same country where so much tragedy had occurred? She is the real champion. The real hero. She has more grit than I could ever have. She doesn't do it because she must. She does it because that is in her nature, her character. That girl I met back in Kentucky in 2008, she is the real deal and I fail constantly to understand why God put that Bluegrass angel into my life.

Josh introduces his son Connor to his Platoon Sergeant,
SFC Christian Lawson, (MAY 2015)

AAR

All training in the military should end with an interactive discussion on what the mission was, the tasks to be trained, and the end state of the training. Known as an After-Action Review or AAR, this venue should help the trained and facilitators see themselves post- action and where they stand going further. Is additional training necessary to obtain the level of proficiency desired? Are they validated in that task whether it be individual or collective? What could they have done better or what resources could be implemented for the next iteration to further improve upon the training?

For these four years of my career, I have done quite a bit of self-reflection on what went wrong and what went right? What are my principles, my non-negotiables if you will, that I retain in my kit bag to facilitate my development and growth. While every man or woman has their own set of beliefs and patterns, I would like to share mine with those young and future officers and all of the Service Members in our force. Please keep in mind, this is where I saw myself as a brand-new minted Captain just leaving the platoon echelon and preparing for the rigors of command.

As Major Robert Rogers had his standing orders for the Rangers, I have my own set of ethos and values that I believe represents some solid truths about our profession. As you, the reader, are no doubt weary after reaching this final stage, I will be brief.

WITHOUT RESILIENCY, STRENGTH AND INTELLIGENCE FALTER IN THE BREACH.

Time and time again, I have been challenged mentally, physically, and spiritually to my limits. There is one absolute truth for every human; there will be setbacks. Loss, heartbreak, injury, failure, or rejection can affect anyone at any time. Over the years, I have experienced them all, some more than once. Life isn't fair and it never will be is a constant I have endured. Being a doctor, Airborne Ranger, professional athlete, or scientist are all outstanding professions but without the ability to get back up after being knocked known, you will never know your true metal. Self-pity is your biggest enemy on this earth. Banish any thoughts from your mind that you cannot accomplish something. Try again. And again. And Again. But never quit on yourself. Because at the end of the day, you alone possess the intestinal fortitude necessary to overcome the obstacles life throws at you. Being well prepared will help you succeed, but your resiliency will keep you from ever accepting failure as an option.

PUSH YOURSELF TO THE THRESHOLD OF FAILURE.

Everyone is guilty of complacency, even me. While I enjoy the beauty of the outdoors, I am guilty of maintaining a lust for air-conditioned bedrooms and a few vices to ease the mind (have you ever tried Woodford Reserve with a splash of Ale-8-One on the rocks?) Being comfortable is no sin by any means, but I inquire to you as I ask myself every day if what I am doing is really challenging me. As an Infantry Officer, I ask myself this question on most days, "What am I doing to prepare myself for the hardest day of ground combat?" As a young LT, this meant going to the field and getting back to the basics in Soldier skills. However, just knowing simple tasks, while essential, will not help you see just how much you can do. As the aforementioned tenant enables you from ever accepting failure, I want you to challenge yourself until you reach the breaking point.

Known as the threshold of failure, I want you to execute where you are most comfortable (60-70%) and then ratchet it up about 10-20% more. This does not always have to be physical. Whether that be as simple as setting the difficulty settings on your online chess game from beginner to intermediate, running your first marathon, to climbing a damn mountain, make the deliberate decision to branch out from the comfort zone and strive to succeed at levels you had not previously performed in. Think of it in terms of strength conditioning. Once you can bench press 225 pounds for ten repetitions, add more weight. Try 275 pounds now and push until you can hit those ten rep sets. Then add more. Bottom line up front: When you are comfortable, you are no longer challenging yourself and striving to improve. Let's conquer complacency.

IT WILL BE THERE TOMORROW.

I am immensely guilty, as you have read, of devoting my soul to my occupation at the expense of my family. When I could have been home by 1700 hours, I stayed many nights past 2000 and even 2100 on occasion. Now, there is a time and place for grinding it out late. There will be short suspense's or fires to put out, even in the healthiest of command climates. That's life. But to make your profession a priority over family and home life runs a fine line that often leads to tracer burn-out or heartbreak. When the opportunity allows, I ask you to disconnect or log out, and go home. Recharge your mind on the commute home and devote yourself to your family or that special hobby that is your passion.

The Army goes rolling along as the song goes, and that email you plan on sending at 1900 hours will probably not be read and definitely not actioned prior to you showing up the next morning. For those of you with families, I strongly encourage that you make a dedicated effort to break away every day at a set time and spend those few precious moments with them. It may not happen every day (as I mentioned due to those fires) but make a consistent disciplined effort to unplug from the cubicle or office and return the favor.

Trust me on this one, the military will get their time. At some point you will retire or leave the service. At the end of that departure ceremony, with a service award pinned to your uniform, you will be left standing there. A janitor is going to come sweeping into the room and advise you that the next ceremony will begin soon and that he needs the facility. And then you leave and take off the uniform for the last time. For the love of God, ensure it is your loved ones who are urging you and not that janitor. The Army will not always be there. Ensure you do whatever you can, whenever you can, to have your family and friends still at your side when you cross the finish line.

DON'T SHY AWAY FROM THE TRANSITIONS. THEY WILL HAPPEN REGARDLESS.

The military is a unique profession where Service Members can become masters in a particular skillset, whether that be small arms, airborne operations, driving tanks, and diving. If you have a hobby that you are passionate about, odds are high that you can get paid to do that hobby for a living in the military. Now, for my officer friends, particularly my combat arms family, this isn't really the case is it? About the only thing I have mastered over the years is PowerPoint and Microsoft Excel, and that's a very soft mastering. Once you know the ropes of a new assignment and you have your rhythm, odds are that the higher ups are about to move you somewhere else to fill a vacancy.

For most of us, being in a duty position can span from a short stint of six months to as long as two years. Thus, becoming a master in anything in such a short period of time with the competing requirements of life is simply unfeasible.

We need well rounded leaders who understand how their particular service works in a holistic fashion. I was kicking and screaming when I was advised to leave the airborne community and try my hand elsewhere in a mechanized or Stryker Infantry Brigade Combat Team to further develop.

You will not be a Platoon Leader for your entire time in the service barring a few exceptions. You will be on staffs rowing away. You will be Company Executive Officers. You will be liaisons. You may even be asked to perform additional duties that don't even correlate to the job description you originally signed up for. Don't pass up the opportunity to try something different. It will be out of your comfort zone. It may even scare you, but isn't that part of the journey? Never pass up a chance to broaden yourself because, down the road, your experience will make you a key agent in the room whom everyone wants at the table.

READ. BE A LIFELONG LEARNER.

Once in a while, open a book and immerse your mind into it. For career officers, this is one of the most valuable pieces of advice I can give you. Professional occupations such as medical doctors, lawyers, and professors are always reading. It didn't stop once they received their degree. New techniques in industry, advanced methods of application, and case studies of successful and failed practices are continuously being shown to facilitate a field's continual improvement over time. Don't fall behind while resting on your laurels. Research and study your particular skillset where you can to further develop your baseline. Then, I want you to devote yourself to what the other professionals are broadcasting, whether that be podcasts, articles, or audiobooks.

I can promise you one thing, nobody ever regrets preparing for a test.

But you are always being tested in your field. And while one "I don't know" wont discredit you, failure in a task that should be elementary to a professional can be career suicide. As we don't grow complacent, but strive to improve on how we operate, nourish your mind by opening or turning on that resource. Take it all in, and then share it with everyone else.

WHAT ARE YOUR PRINCIPLES?

Some things for me are non-negotiables. Period. My wife is my number one priority. No alcohol on Sunday prior to noon. Self-improve in everything. My faith in Christ is my own business and not anyone else's. Respect all people, their beliefs, and their cultures, even the ones I disagree with. Finally, I can accomplish anything I set my mind to. These are my core tenants that I refuse to compromise to anyone, regardless of the occasion.

What principles I keep are different from everyone else's. Some people don't eat meat. Some worship in ways that seem alien to me. Some people want to end carbon emissions. While I don't have to agree with other's beliefs, I still strive to tolerate them until it poses a detriment to my family or myself. Either way, create a list of some non-negotiables you believe in and evaluate them. Odds are, they are completely ethical, moral, and legal. Those that blur the lines, take a hard look at yourself, and seek that self-reflection.

Knowing yourself is critical to self-awareness and understanding where you have room for improvement and what to maintain. I have other tangibles such as exploring and fitness, but they conflict with the other non-negotiables. Therefore, they did not make the list. Empower yourself with a core set of beliefs that you refuse to negotiate on. If you need assistance, consult a unit Chaplain who I am sure can assist you. At the end of the day, what defines the man or woman that you are?

GET YOUR HANDS DIRTY.

For you young officers, it is all too tempting to remain in that air-conditioned cubicle, office, or tent. You have an endless number of tasks that never seem to shorten. As 1SG Anspach added more tasks to my plate, I felt myself increasingly on email, writing memorandums, or attending ad-hoc meetings. The Service Members under your charge know you are busy. However, it is usually not expected of you to perform menial tasks such as pushing a broom, taking out the trash, or mowing the grass in front of the Battalion Headquarters. Here is my challenge to you. Take out your Common Access Card, unplug for a moment, and go out of your office and just observe. You will probably see your team doing things that they didn't expect to do in the recruiting commercial. Those menial tasks, while essential, are always left to them.

Pick up a broom and sweep. At the bare minimum take the trash out of your own workspace and clean the floor that your dirty boots contaminate with every step. Volunteer yourself to serve chow in the field as you shouldn't be eating first anyways, right? Do whatever menial tasks you could do to make the team's life easier. And if you physically can't do it, use that powerful college education to come up with a unique way to get it done. Here is an example:

I think at one point the First Platoon boys had to mow a giant section of the fishbowl during Operation Clean Sweep (it's a real event) in June of 2014. That was their task for the next couple of days and all we had was some old push mowers. I left my office and went outside to see what I could do. That's when I saw the civilian contractors mowing the residential area on those sweet zero turn mowers. I walked over to one of them and promised a log of Copenhagen and six pack if they could mow our section. Without hesitation, on that hot ass day, he did. And what would have taken the boys a long time in the sun to do was complete in less than an hour.

DEVELOP YOUR TEAM AND CULTURE.

From fire teams and staff sections to the division echelon, we all work towards a common end state. Rarely will you be asked to do anything by yourself in the military. Thankfully, the task organization is in place to assist you in mission accomplishment. Whether a leader or follower, what is the climate where you operate? Would you want to be the 1SG or Commander of this outfit one day? Would you want the Service Members in this organization on your team? Hopefully the answer is yes.

CPT Blackburn created an outstanding command climate and team that was known as Brutal Nation. Everyone wanted to be there, and the retention rates showed from that era. Building a command climate of this level took an enormous amount of work and time. It just didn't happen overnight. Most importantly, he didn't do this with flashy speeches and gimmicks. He accomplished his legacy by how much he showed he cared. He and 1SG Goodart took every opportunity to look out for us, develop us as professional leaders and Soldiers, and prioritized hard, realistic training. Everything was as transparent as it could get. Soldiers weren't kept in the dark and we knew our schedule months out, by the day. And he was always out there, getting dirty with his Paratroopers and embracing the suck.

While you still may have time before becoming a Company Commander, odds are a platoon is in the cards for your near future. Work on what culture you want to create as you build your team. Your Platoon Sergeant and you should be shoulder to shoulder in preparing for the toughest day of ground combat. Train realistically and try your hardest to protect your team's time. Learn about every member of your unit. Everyone has a stake in your organization and your failures or successes impact each and every single person. "Don't coddle Joe," as SSG Lawson always reminded me, but strive to empathize with them on their down days, do everything in your power to ensure they are taken care of administratively, and challenge them to be bold. The climate in your platoon is the first litmus test until your time to take the guidon. Don't rush it and try to learn from the experience.

WAR IS HELL.

Time and time again we tell ourselves that we are members of the greatest fighting force in modern history. That nobody can reach out and touch us and anybody foolish enough to try will get what's coming to them. Well, I hate to break it to you, but some dusty Pashtun hashish farmer blew off my leg after probably being paid $10 by the Taliban to bury an IED. I had the best training there is to offer and was equipped with technologies that he couldn't even fathom. But on April 15, 2012, the day was his.

I completely understand the alure of deploying to combat. You will never feel closer to another human being than when suffering together in the most inhospitable and austere conditions imaginable. Hollywood is quick to glamorize fire fights and war. Firefights typically are short in my experiences, ranging from a few seconds to a few minutes. The adrenaline rush is awesome once they begin, but deep down, you just can't wait for it to be over. This was true during much of the GWOT era, barring a few exceptional cases that made headlines. The enemy simply couldn't match us in a gun fight and remained in the shadows for most of the time.

Today is a different world though. We have adversaries and near peer competitors that have now caught up in the technological race. Perhaps most disturbing though is that while we were in the graveyard of empires, they studied us and took notes. They saw what the American public couldn't stomach and documented where our shortcomings were. Then, they turned out new doctrine to capitalize on this. Do not underestimate your opponent in the next round. Odds are, they are just as capable if not better, than you.

I volunteered for the 4th Brigade Combat Team in 2011 so that I could embark on my first deployment to combat as a Platoon Leader. While I do not regret the decision, it cost me severely. I was cocky and thought nothing could touch me and I paid dearly for my hubris. You are not immune once the hardware flies or the pressure plate is detonated.

There is a dichotomy of me saying war is hell. I am, after all, a professional Soldier whose sole purpose is to enable our mission of fighting and winning our nation's wars on the land. But one thing I have learned is that there is nothing glorious about combat. It is a crude endeavor where the metal meets the flesh and is the scariest entity imaginable. I just ask that you do everything in your power to meet that moment with a bit of humility.

HAVE FUN AND ENJOY THE RIDE.

Finally, as the title states, try to enjoy yourself. I have made some outstanding friends, visited some exotic places, and learned more about myself in a few years than most could in a lifetime. The military took care of my medical bills, cost of living, housing, and ensured I would receive a paycheck. The best part was, I never went to bed wondering if I could feed my family next week or if I was going to be laid off.

With those dynamics that millions of Americans worry about vanquished, I was free to focus on what mattered. I took up traveling all over the world. I poured myself into my studies to obtain an advanced degree. And I now have an amazing family that makes me the happiest man on earth. I jumped out of airplanes, forded snake infested swamps, snowboarded in the Rockies, hiked on the Appalachian Trail, and drank cheap beer on exotic beaches. Life has been too kind to me because I have learned to never say no to new adventures and experiences where I could further grow as a human being.

When riding in that tank turret, or viewing the sky from the paratroop door, take a moment and reflect on how you got here. Know that some poor citizen in a cubicle with no windows envies you. As you ponder, try to realize that you have it made. You are a hero to 99% of the country that chose their respectable occupations outside of service. That is something to be proud of. Finally, when you get done with that field operation, and you are dirty, tired, wet, hot, and hungry, look out over that horizon when the sun sets and try to smile. Because at the end of it all, you are living your best days, even if you don't know it yet.

EPILOGUE

As of this writing, the war in Afghanistan is over. Victory was not obtained, and the withdrawal drew immense criticisms due to the waste of blood and treasure. Watching the C-17's takeoff with terrified Afghans clinging for life on the fuselage haunts me to this day. A lover of history, the comparison to the withdrawal of the Saigon Embassy before the fall of South Vietnam in April 1975 was so acute.

The most difficult aspect of watching those scenes from my home in Kentucky was wondering not just for the safety of the Afghans who had helped us, but for those brothers and sisters of mine that had served with me. For almost a week, I talked to former members of First Platoon and Brutal Nation who were there. Paratroopers who had left the Army on their terms now struggled to reconcile how this came to pass. Their legacy was not one of victory as we had imagined in 2012 and 2014, but one of tragedy as a two-decade long conflict concluded with the white flag of the Taliban being raised throughout the bases we once occupied.

For hours, I talked to those men. Many of them inebriated, many of them crying, and some on the verge of ending their lives. I had to explain to them that they were not at fault. That this chapter in their lives was one of honor and not wasted because of the way the campaign ended. Every single one of us was there as a volunteer and once we went out that gate to patrol, we were there for each other. Soldiers, Sailors, Marines, and Airmen fought like hell in places that no longer register to the public, but to us, have become shrines due to the amount of blood, sweat, and tears we poured into them.

But at the end of the day, our legacy is not determined by the whims of politicians or diplomats. It is decided by our own actions and character. The Service Members who fought in the Global War on Terror are an example to all Americans during a time of uncertainty and bipartisanship. Republican, Democrat, white, black, Hispanic, Christian, Muslim, Jewish, gay, straight, young, and old. Some of them came from working class families that could barely feed themselves. Others came from wealthy backgrounds looking for adventure. Every single one of them was different. But there is an absolute truth. And that is that no matter how different someone is in character or appearance, every single one of these Service Members would have fought for their brother and sister and laid down their life without hesitation.

Part I of this chronicle is for these Service Members who answered the call and put their lives on hold. Afghanistan to this day holds memories for me. As my friend CSM Brian Disque cautioned me as a Cadet in 2010, "Afghanistan will make you wish you had never been born." He was right to an extent. That country cost me my left leg and much more. It cost me friends. It cost my brother's mental state. But it did not take my soul. The Taliban wounded me severely on April 15, 2012, but also did something none of them could have imagined. They sparked a fire in me that has continued to burn brighter with every day I live. I have become mentally and physically stronger because of the experiences I encountered.

That catalyst in the form of an IED was just a means to test my will, a challenge by God for me to realize just how much I could overcome. And just as I have had to overcome much to get the most out of life, I treasure the knowledge I have to let you in on this little secret. You're the only obstacle in your path. Breach it and move out.

ACKNOWLEDGEMENTS

I did not embark on this effort in a solo capacity. Throughout my career, numerous officers and NCOs facilitated my growth as an Infantry Officer. First, I want to thank Command Sergeants Major Brian Disque, one of my first real mentors since I was a fledgling Cadet. You hated me those first few weeks at LDAC. Don't deny it. To Colonel Marcus Franzen, for guiding me spiritually, mentally, and physically. For challenging me to take the hardest assignments and for shaping my mindset to never surrender to my demons. To Brutal-6 himself, CPT Bryan Blackburn, for showing me what servant leadership truly is and how to lead a bunch of pipe hitters where the metal meets the flesh.

To the organizations who dedicate countless hours and funding towards our deceased and wounded Service Members and families. Specifically, America's Fund and Semper Fi Fund who have been there for my family and I time and time again. To Miracles 4 Melanie which is one of the most selfless organization I have ever had the privilege of championing. Erica and Lou, your devotion is unmatched in all regards.

I want to thank my parents, Randy and Vicki Pitcher, for raising me and putting up with my ego. I love you both. To my father and mother-in-law, Clay and Vicki Smith. Thanks for those second, third, and fourth chances which I never did deserve. The grace you show is an example for us all.

Most importantly though, as if you weren't mentioned enough

throughout this work, my beautiful wife Michelle. She is the epitome of the military spouse who stands behind her Soldier. Without her, I wouldn't be here today. While Bluegrass Grit may recount my experiences, the true grit belongs to her. I love you and should thank God more often for your presence in my life.

www.ingramcontent.com/pod-product-compliance
Lightning Source LLC
Chambersburg PA
CBHW061603120626
46550CB00004B/1593